FOOD MOVEMENTS UNITE!

. . .

STRATEGIES TO TRANSFORM OUR FOOD SYSTEMS

EDITED BY ERIC HOLT-GIMÉNEZ

~•~

FOOD FIRST BOOKS
OAKLAND, CALIFORNIA
www.foodfirst.org
www.foodmovementsunite.org

Food First Books, 398 60th Street, Oakland, CA 94618-1212
(510) 654-4400 www.foodfirst.org

Cover, web, and text design by Richard Jonasse,
Birdseye Studio, Oakland, CA

Development Editor: Patrick Koohafkan

Copy Editor: Carrie Laing-Pickett

Proofreader: Martha Katigbak-Fernandez

Library of Congress Cataloging-in-Publication Data:
Food movements unite! : strategies to transform our food systems / edited by Eric Holt-Giménez.

p. cm.\

ISBN 978-0-935028-38-6 -- ISBN 978-0-935028-39-3
1. Food supply. 2. Agricultural productivity. 3. Food industry and trade.
4. Food supply--Environmental aspects. 5. Agriculture and state.
I. Holt-Gimenez, Eric.

HD9000.5.F5954 2011
338.1'9--dc232011033220

Food Movements Unite! : Strategies to transform our food systems / edited by Eric Holt-Giménez

Food First Books are distributed by: Perseus Distribution
387 Park Ave South, 12th Flr
New York, NY 10016
Tel: 212-340-8118
http://www.perseusdistribution.com/

Printed in Canada.

WHY HUNGER: A leader in building the movement to end hunger and poverty by connecting people to nutritious, affordable food and by supporting grassroots solitions that inspire self-reliance and community empowerment.

http://www.whyhunger.org/

Roots of Change

Roots of Change brings a diverse range of Californians to the table to build a common interest in food and farming so that every aspect of our food—from the time it's grown to the time it's eaten— can be healthy, safe, profitable, affordable and fair.

http://rootsofchange.org/

More and Better has been established in 2003 to join and sustain the fight to eradicate hunger an-dpoverty. It is an International network embracing social movements, civil society (CSO), non governmental organizations (NGO) and a core of national unified campaigns from all over the world.

http://www.moreandbetter.org

Grassroots International works around the world to help small farmers and other small producers, indigenous peoples and women win resource rights: the human rights to land, water and food.

http://www.grassrootsonline.org

FOODFIRST
BOOKS

The Institute for Food and Development Policy (Food First) is a "people's think-and-do tank." Our work informs and amplifies the voices of social movements fighting for food justice and food sovereignty.

http://www.foodfirst.org

US FOOD SOVEREIGNTY
ALLIANCE

The US Food Sovereignty Alliance works to end poverty, rebuild local food economies, and assert democratic control over the food system. We believe people have the right to healthy, culturally appropriate food, produced in an ecologically sound manner.

http://www.usfoodsovereigntyalliance.org/

Acknowledgments

This book would not have been written had it not been for the support of the many people who donated their time and expertise to the project because they believed in the importance of building a strong global food movement. Tanya Kerssen helped contact authors and, along with Marilyn Borchardt, Annie Shattuck, Zoe Brent, and Leonor Hurtado helped with the book's conception and provided helpful comments along the way. Bill Wroblewski filmed and edited interviews for the book and the website. Richard Jonasse dedicated many hours to the book design.

We would also like to thank the many generous photographers who put their wonderful images on Creative Commons, including the following, whose work contributed to the cover collages: Meredith Kahn, Schubert Ciencia, SEIU International, Andreas Wilkes (Farming Matters), Free Range Jace, Vredeseilanden, Grant Neufield, Jean-Marc Desfilhes, Luciano Garcia, Kasuga Sho, Daniela Hartman, ~ Abdulhai A. Al-Abdulhai~, International Institute for Tropical Agriculture, Bo Nielsen, Jean Marc Desfilhes, Kris Krug, Brennan Cavanaugh, Teemu Mantenen, fuzheado, International center for tropical agriculture-CIAT, ImageMD, Swissaid, hey.kiddo, Edible Office, and World Agroforestry Centre.

Thanks to Gabriel Holt and Eva Thalia Holt-Rusmore, who spent many hours painstakingly translating chapters from Spanish and Portuguese into English, as well as Deanna Drake Seeba, who translated from French. Thank you Patrick Koohafkan, Carrie Pickett, and Martha Katigbak for your careful editing. Also thanks to Renee MacKillop, Aja Peterson, Celeste Ariel Peifer, Reema Cherian, Stephanie Kennedy, and Rebecca Mistruzzi, Food First interns who worked on sections of the book.

We also want to especially thank the following benefactors who responded to a special appeal and generously donated funds to cover our publishing costs:

Joan and Wise Allen, Wayne Alt, Valerie Anderson, Clifford Anderson, Henry Atkins, Jessica Bader, Catherine Badgley, Gregory Bartha, Jim Berger, Sid Berkowitz, Scott Bohning, Lorraine Bonner, Nicholas Bridge, George Brieger, Steve Buban, Ingrid Buntschuh, Marie Cadenhead, Gerald Cavanaugh, Lynn Cawthra, Carol and Gregory Codner, Bruce Cohen, Mitch Cohen, Dorothy Corbett, Martha Corry, Dr. Monica Courtney, Keith Dangelo, Richard De Wilde, Sylvia Demarest, Maxine and Michael Denniston, Paul and Eleanor Diesing, Uta Dreher, Grace Dumenil, Roger Eberly, Charles Engstrom, Elizabeth Esteves and Saadedine Tebbal, Ron and Bonnie Ettinger, Sabrina and David Falls, Wendy Fassett, Abraham Flaxman, Susan Foerster, Pat Foley, Hunter Francis, Eva Fuld, James and Virginia Gailey, Jan Garrett, LeRoy Gaskin, Audrey Gilmore, Daniel Gladstone, John Gloor, Raymond Goggan, Jean Gore, Margaret Grip, Linda Grove, Victoria Hadd-Wissler, Darcy Hamilton, David and Eleanor Hammer, Margaret Hammond, Hildegarde Hannum, Evan Hartman, Gail Hartman, Everett Heath, Louis Hellwig, Narada and Kali Hess, Lois and Ron Hines, William Horton, Celeste Howard, Anne Howe, Helen Jacobsen, Mary Jean Jawetz, Mark Jay and Karen Pakula, Laura Juraska, David Kaun, Vivienne and Raymond Kell, Elaine Kihara and David Sweet, Robert and Rita Kimber, Gregory King, Gloria and F. Ross Kinsler, John Klingman, Kathryn Kram, Kallyn Krash and John Emmanuel, Kathy Labriola and Rick Lewis, Miriam Laster, Eileen and Paul LeFort, Peter Linenthal, David Loeb, Betty Lucas, William Lupatkin, Glenn Lyons and Nancy Gerdt, Joseph Mabel, Locke McCorkle, Sarah Metzger, M. Jane Meyerding, Ron Miller, Beth Moore, Lois Moschella, John and Hatsumi Moss, Mary Myers, Abby Nash, David Nelson, Elaine Netherton, Lane Nevares, William Nisbet, Meg Palley, Barbara Parsons, H. F. W. Perk, Linda Peterson, James Reilly, Barbara and Brian Robinson, Peter Robrish, Kathryn Rodriguez, Lorraine Rogers,

Gordon Rogoff, Norma Rossel, Mary Rowe, Dan Ruesnik, Rachel and Joel Samoff, Janis Sarles, Ruth Schultz, Elizabeth Schwerer, Corinne and David Scott, Dan Seligson, William Sewell, Peter and Elizabeth Shepherd, John Simmons, Tony Stern, Caryl and Milton Stone, Shirley and Harold Strom, Heather Sullivan-Catlin, Barb Swanson, Daniel Szyld, Bahram Tavakolian, Eugene Teselle, Murray Tobak, Joseph Towle, Richard and Cathleen Vaught, Erika Voss, Victor Wallis, Jane Wentworth, Thurman Wenzl, John and Carolyn White, Sharon Whitmore, Patricia and Thomas Willis, Jane Wood, Helen Yost, Barbara Young, Lawrence Young, Robert Zevin, and Charles Zug.

Finally, we are profoundly grateful to the hundreds of thousands of youth, women, and men—the advocates and practitioners of the food justice and food sovereignty movements—who are on the front lines in the struggle to transform our food systems. They are the inspiration for this book. For them, giving up hope is not an option.

Table of Contents

Preface

FOOD SOVEREIGNTY:
A STRUGGLE FOR CONVERGENCE IN DIVERSITY

by Samir Amin

Family Farms, Modern Agriculture and the Production of Hunger

MODERN FAMILY AGRICULTURE in western Europe and the United States is highly labor productive. Producing 1,000 to 2,000 tons of cereal equivalents annually per worker, it has no equal and has enabled less than five percent of the population to supply whole countries abundantly and to produce exportable surpluses. Though it is not necessarily the most productive form of agriculture measured in tons per hectare (ha), modern family farming has an exceptional capacity for absorbing innovations and adapting to both environmental conditions and market demand.

Though deeply embedded in capitalism, family agriculture is different from industrial agriculture in that it does not share that specific characteristic of capitalist production: industrially organized labor. In the factory, the number of workers enables an advanced division of labor, which is at the origin of the modern leap in productivity. On family farms, labor supply is reduced to one or two individuals (the farming couple), sometimes helped by one, two, or three family members and associates or permanent laborers, and also, in certain cases, a larger number of seasonal workers (particularly for the harvesting of fruit and vegetables). Generally speaking there is not a definitively fixed division of labor, the tasks being complex, polyvalent, and variable. In this sense, family farming is not capitalist. Nevertheless, modern family agriculture in the Global North is an inseparable, integrated part of the capitalist economy, and its combined productivity and labor efficiency bring tremendous productivity and resiliency to the global agrifoods system.

The labor efficiency of the modern family farm is due primarily to its modern equipment and to the fact that it possesses 90% of the tractors and agricultural equipment in use in the world. In the logic of capitalism, farmers are both workers and capitalists, and their income should correspond to the sum of their wages for work and profit from ownership of the capital being used. But it is not so. The net income of farmers is comparable to the average (low) wage earned in industry in the same country. State intervention and regulatory policies in Europe and the United States favoring overproduction (followed by subsidies) ensure that profits are collected not by the farmers but by segments of industrial, financial, and commercial capital further up and down the food-value chain.

Despite its efficiency, the agricultural family unit is only a subcontractor caught in the vise between upstream agroindustry (which imposes GMOs and supplies the equipment and chemical products) and finance (which provides the necessary credits), and downstream by the traders, processors, and commercial supermarkets. Self-consumption has become practically irrelevant to the business of family farming because the family economy depends entirely on its market production. Thus, the logic that commands the family's production options is no longer the same as that of the agricultural peasants of yesterday or of today's Third World countries. Because of their absolute subjugation to market forces, family farmers are victims of capitalism's system of mass production, both as producers and consumers. This reality links them to peasant producers in the Global South and to the growing underclass of consumers of "mass food" produced by the corporate agrifood industry worldwide.

The Third World counterparts of Northern family farmers are the peasant cultivators that constitute over a third of humanity—two and a half billion human beings. The types of agriculture vary, from the unmechanized use of so-called green revolution products (fertilizers, pesticides, and hybrid seeds), the production from which has risen to 100–500 quintals per laborer; to those caught in the green revolution's negative spiral of "involution," whose production has dropped to around 10 quintals per laborer and continues falling, despite costly increases in inputs. Another, growing category of productive peasant farmers are the "agroecological" producers managing farm and watershed-scale ecosystem functions to maintain productivity and resilience, and to lower their production costs, and whose productivity—when measured in kilos per ha—rivals both industrial and family farming. Nonetheless, the gap between the average production of a farmer in the North and that of Southern peasant agriculture, which was 10 to 1 before 1940, is now 100 to 1. In other words, the rate of progress in agricultural productivity has largely outstripped that of other activities that, when combined with global overproduction, bring about a drop in real price from 5 to 1.

The peasant (family) agriculture in the countries of the Global South, like its Northern counterpart, is also well integrated into world capitalism. However, closer study immediately reveals both the convergences and differences in these two types of "family" economy. There are huge differences, which are visible and undeniable: the importance of subsistence food for survival in the peasant economies; the low labor efficiency of this nonmechanized agriculture; the impossibly small land parcels and their systematic dispossession and destruction by urbanization, agrofuels, and industrial agriculture; the vast poverty (three-quarters of the victims of global undernourishment are rural); and the sheer immensity of the agrarian problem (the peasantry is not a 2%–5% sector of a larger, industrialized society, but makes up nearly half of humanity).

In spite of these differences, peasant agriculture is part of the dominant global capitalist system. Peasants often depend on purchased inputs and are increasingly preyed on by the oligopolies that sell them. Further, these farmers feed nearly half of the world's population (including themselves). For green revolution farmers (approximately half of the peasantry of the South), the siphoning off of profits by dominant capital is severe, keeping them desperately poor (as evidenced by the epidemic of bankruptcies and farmer suicides in India, for example). The other half of the peasantry of the South, despite the weakness of its production, has a combined annual income of US$2.3 trillion and is growing at a rate of 8% a year (and is therefore seen as a $US1.3 trillion/year potential market).

The Industrial Colonization of Peasant-Based and Family Food Systems

In response to the global food crisis, capital's corporate food regime—made up of Northern governments, multilateral institutions, agrifood oligopolies, and big philanthropy capital—propose using public tax revenues to modernize areas in the Global South of high agricultural potential (i.e., "breadbasket" regions with good land and access to irrigation) to bring them into global markets. This, we are invited to believe, will eradicate rural poverty and lead to national economic development for poor countries in the Third World, thus bringing an end to world hunger.

This strategy is supported by the "absolute and superior rationale" of economic management based on the private and exclusive ownership of the means of production. According to conventional economics, the unregulated market (i.e., the transferability of ownership of capital, land, and labor), determines the optimal use of these factors of production. According to this principle, land and labor become merchandise like any other commodity, and are transferable at the market price in order to guarantee their best use for their owners and for society as a whole. This is nothing but a mere tautology, yet it is the one upon which all critical economic discourse is based.

The global system of private land ownership required for the free movement (and concentration) of capital is justified in social terms by the argument that private property alone guarantees that the farmer will not suddenly be dispossessed of the fruit of his or her labor. Obviously, for most of the world's farmers, this is not the case. Other forms of land use can ensure that farmers (as well as workers and consumers) equitably benefit from production, but the private-property discourse uses the conclusions that it sees fit in order to propose them as the only possible "rules" for the advancement of all people. To subjugate land, labor, and consumption everywhere to private property as currently practiced in capitalist centers is to spread the policy of monopoly "enclosures" the world over, to hasten the dispossession of peasants, and to ensure the food insecurity of vast poor communities. This course of action is not new; it began during the global expansion of capitalism in the context of colonial systems. What current dominant discourse understands by "reform of the land tenure system" and "new investments in agriculture" is quite the opposite from what the construction of a real alternative based on a prosperous peasant economy requires. This discourse, promoted by the propaganda instruments of collective imperialism—the World Bank, numerous cooperation agencies, and also a growing number of nongovernmental organizations (NGOs), with financial backing from governments and philanthropy capital—understands land reform to mean the acceleration of the privatization of land and nothing more. The aim is clear: create the conditions that would allow modern islands of agribusiness to take possession of the land it needs in order to expand.

But is the North's capitalist modernization of the South's agriculture really desirable? Is it even possible?

One could easily imagine that by concentrating the production of 2.5 billion people onto 50 million new modern farms on large areas of prime agricultural land with access to subsidized credit, these farms could certainly produce that half of the world's food currently obtained from peasant agriculture. Perhaps this might even free up some of the estimated 276 million ha needed to meet the North's agrofuel demands (though it might not provide enough water). But what would happen to the livelihoods and food systems of the billions of "noncompetitive" peasant producers? They would be inexorably pushed off the land and eliminated in a short period of time, a few decades at most. What would happen to these billions of human beings who, for better or for worse, have the capacity to feed themselves? Within a time horizon of 50 years, no marginally competitive industrial development—even under a far-fetched hypothetical scenario of 7% yearly growth, could begin to absorb even a third of this massive labor reserve. They will be condemned to hunger, migration, and suffering, not because of a lack of food, but because they will be forced off the land and into a dysfunctional food system that keeps them in intractable poverty and food insecurity.

Capitalism, by its nature, cannot solve the global hunger crisis because it can't resolve the historical agrarian question of how to mobilize the surplus from peasant agriculture to industry without eliminating that same peasantry from agriculture. Although capitalism did accomplish this transition for the industrial societies of the Global North, this proposition does not hold true for the 85% of world population in the Global South. Capitalism has now reached a stage where its continued expansion requires the implementation of enclosure policies on a world scale similar to those at the beginning of capitalist development in England, except that today the destruction on a world scale of the "peasant reserves" of cheap labor will be nothing less than synonymous with the genocide of one-third to one-half of humanity. On one hand the destruction of the peasant societies of Asia, Africa, and Latin America; on the other, trillions in windfall profits for global capital derived from a socially useless production unable to cover the needs of billions of hungry people in the South, while increasing the number of sick and obese people in the North.

We have reached the point at which in order to open up a new area for capital expansion it is necessary to destroy entire societies. Imagine 50 million new "efficient" modern farms (200 million human beings) on one hand, and over two billion excluded people on the other. The profitable aspect of this capitalist transition would be a pitiful drop of water in a vast ocean of destruction. The effect of increased out-migration from the countryside will shift capital's social misery to new and existing urban communities of poor and underserved "surplus people." The breakdown of the global food system reflects the fact that, despite its neoliberal bravado, capitalism has entered into its phase of senility because the logic of the system is no longer able to ensure the simple survival of humanity. Capitalism's continued expansion into Southern agriculture will result in a planet full of hungry slums. Once a creative force sweeping away the bonds of feudalism, capitalism has now become barbaric, leading directly to genocide. It is necessary to replace it—now more than ever before—by other development logics that are more rational and humane.

What is to be done? Different movement leaders address this historical question in different ways within this volume. I wish to address it—as they do—without falling into past or modernist romanticisms but by advancing a new vision of food sovereignty.

There Is No Alternative to Food Sovereignty

Resistance by the peasants, small family farmers, and poor consumers most affected by the dysfunctional global food system is essential in order to build a real and genuinely humane alternative. We must ensure the functionality and resilience of family and peasant agriculture for the visible future of the 21st century because this allows us to resolve the agrarian question underlying world hunger and poverty. Peasant, family, and improved, agroecological agriculture—along with a new relation with consumers and labor—are essential to overcome the destructive logic of capitalism.

I personally believe this operation will entail a long, secular transition to socialism. The initial weight of this transition will be primarily in the South, but will also need to address both rural and urban food systems in the North. We need to work out regulatory policies for new relationships between the market and family agriculture, between producers and consumers, between the North and the South, and between the rural and urban.

This is a historically large, multifaceted task that must address the structural rules governing capitalist food systems. To begin with, the agenda of the World Trade Organization and its attendant global-market model must quite simply be refused. At the national, regional, and subregional levels, regulations adapted to local food systems must protect national, smallholder production and ensure food sovereignty—in other words, the delinking of internal food prices and the rents (profits) of the food-value chain from those of the so-called world market.

A gradual increase in the productivity of peasant agriculture based on different combinations of agroecological and input-mediated strategies will doubtless be slow but continuous, and would make it possible to control the exodus of the rural populations to urban areas (in the North and South) and provide opportunities to construct mutually beneficial, autonomous food systems in underserved communities in regards to local economies, food supply, and diet. At the level of what is called the world market, the desirable regulation can probably be done through interregional and rural-urban agreements that meet the requirements of a kind of sustainable development that integrates people rather than excludes them.

At global scales, food consumption is assured (through competition for 85% of it) by local production. Nevertheless this production corresponds to very different levels of satisfaction of food needs: generally good for North America and western and central Europe, acceptable in China, mediocre for the rest of Asia and Latin America, disastrous for Africa. The United States and Europe have understood the importance of national food sovereignty very well and have successfully implemented it by systematic economic policies. But, apparently, what is good for them is not so for the others! The World Bank, the Organisation for Economic Co-operation and Development (OECD), and the European Union try to impose an alternative, which is "food security." (A similar prescription is applied by national governments to the slums of the Global North, where the food security of low-income communities is achieved through the industrial production of low-quality "mass food.") According to them, the Third World countries do not need food sovereignty and should rely on industrial agriculture, mass food, and international trade to cover the deficit—however large—in their food requirements. This may seem easy for those countries that are large exporters of natural resources like oil or uranium, or to affluent consumers who can afford to eat outside the circuits of mass consumption. For the others, the advice of the Western powers is to maximize specialization of agricultural commodities for export, such as cotton, tropical drinks, oils, and, increasingly, agrofuels. The defenders of "food security" for others—not for themselves—do not consider the fact that this specialization, which has been practiced since colonization, has not improved the miserable food rations of the peoples concerned and has resulted in a global epidemic of diet-related diseases.

On top of this, the economic crisis initiated by the financial collapse of 2008 is already aggravating the situation and will continue to do so. It is sad to note how, at the very moment when the crisis illustrates the failure of the so-called food security policies, the partners of the OECD cling to them. It is not that government leaders do not understand the problem. This would be to deny them the intelligence that they certainly possess. But we cannot dismiss the hypothesis that "food insecurity" is a consciously adopted objective and that food is being used as a weapon. Without food sovereignty, no political sovereignty is possible. Without food sovereignty, no sustainable food security or food justice—national or local—is possible.

While there is no alternative to food sovereignty, its efficient implementation does in fact require the commitment to the construction of deeply diversified economies in terms of production, processing, manufacture, and distribution.

New peasant organizations exist in Asia, Africa, and Latin America that support the current visible struggles. In Europe and the US, farmer, worker, and consumer organizations are forming alliances for more equitable and sustainable food systems. Often, when political systems make it impossible for formal organizations to form (or have any significant impact), social struggles take the form of "movements" with no apparent direction. Where they do exist, these actions and programs must be more closely examined. What social forces do they represent, whose interests do they defend? How do they struggle to find their place under the expansion of dominant global capitalism?

We should be wary of over-hasty replies to these complex and difficult questions. We should not condemn or dismiss many organizations and movements under the pretext that they do not have the support of the majority of peasants or consumers for their radical programs. That would be to ignore the formation of large alliances and strategies in stages. Neither should we subscribe to the discourse of "naive alter-globalism" that often sets the tone of forums, and fuels the illusion that the world would be set on the right track only by the work of disperse social movements.

Convergence in Diversity

Whether it was due to growing pauperization, growing inequality, growing unemployment, or growing precariousness, it's only normal that people started resisting, protesting, and organizing around the world. People are struggling for rights, for justice. Social movements are by and large still on the defensive, facing the offensive of capital to dismantle whatever they had conquered in the previous decades, trying to maintain whatever could be maintained. But even if perfectly legitimate social movements of protest are growing everywhere, they remain extremely fragmented. What is needed is to move beyond fragmentation and beyond a defensive position into building a wide progressive alliance emboldened by the force of a positive alternative.

The balance of forces cannot be changed unless these fragmented movements—such as the movements for food sovereignty, food justice, and food democracy—forge a common platform based on some common grounds. I call this "convergence in diversity": that is, recognizing the diversity, not only of movements, which are fragmented, but of the political forces that are operating with them, of the ideologies and even visions of the future of those political forces. This has to be accepted and respected. We are not in a situation in which a leading party alone can create a common front. It's very difficult to build convergence in diversity, but unless this is achieved, I don't think the balance of forces will shift in favor of the popular classes.

There is no blueprint for convergence in diversity. Forms of organization and action are always invented by the people in struggle—not preconceived by some intellectuals to then be put into practice by people. If we look at the previous long crisis of capitalism in the 20th century, people invented efficient ways of organizing and of acting that worked well at the time: for example, trade unions, political parties, and wars of national liberation all produced gigantic progressive change in the history of humankind. But they have all run out of steam because the system has itself changed and moved into a new phase. And now, as Antonio Gramsci wrote, the first wave has come to an end. The second wave of action to change the system is just starting. The night has not yet completely disappeared; the day has not yet completely appeared, and in this crisis there are still a lot of monsters in the shadows. To move from that fragmented and defensive position into some kind of unity and to build convergence respecting diversity with strategic targets requires the *repoliticization* of the social movements. Social movements have chosen to be depoliticized because the old politics—the politics of the first wave—has come to an end. It is now up to the social movements to create new forms of politicization. It is the responsibility first of activists in the grassroots movements to see that however legitimate their action, its efficiency is presently limited by the fact that it doesn't move beyond a fragmented struggle. But it is also the responsibility of the intellectuals—not the academics, but those thinkers and political people operating in politics—to consider that there is no possibility of changing the balance of power without joining the struggle being carried forward by the social movements—not to dominate them or seek their own fame, but to integrate the activity of grassroots social movements into their political thinking and strategies of change. The activist-intellectuals in this book have taken up both of these challenges. We would all do well to follow their example.

Introduction

FOOD MOVEMENTS UNITE!
STRATEGIES TO TRANSFORM OUR FOOD SYSTEMS

By Eric Holt-Giménez,
Executive Director, Food First

THE CORPORATE FOOD REGIME dominating our planet's food systems is environmentally destructive, financially volatile, and socially unjust. Its central role in creating the global food crisis is well documented. Unfortunately, the "solutions" advanced by our national and global institutions call for developing more of the same destructive technologies, global markets, and unregulated corporate power that brought us the food crisis in the first place. We need a vision for real solutions—not from the actors and institutions that are causing the problems, but from those who are most affected by the poverty and hunger that the corporate food regime produces.

This book is not a critique of the corporate food regime; it is a window into the thinking and actions of the social movements fighting to bring our food systems under democratic control. It is about the emergence of alliances for the transformation of our food systems.

A dynamic global food movement has risen up to confront the corporate assault on our food. Around the world, local food justice activists have taken back pieces of their food systems through local gardening, organic farming, community-supported agriculture, farmers' markets, and locally owned processing and retail operations. Food sovereignty advocates have organized for land reform, the end of destructive global-trade agreements, and support for family farmers, women, and peasants. Protests against—and viable alternatives to—the expansion of genetically modified organisms (GMOs), agrofuels, "land grabs," and the oligopolistic control of our food are growing everywhere every day, "breaking through the asphalt" of a reified corporate food regime with a vision of hope, equity, and sustainability.

The social and political convergence of the "practitioners" and "advocates" in these food movements is well underway, as evidenced by the growing trend in local/regional food policy councils in the US; the coalitions for food sovereignty spreading across Latin America, Africa, Asia, and Europe; and the increasing attention to practical, political solutions to the food crisis appearing in academic literature and the popular media. The global food movement springs from strong commitments to food justice, food democracy, and food sovereignty on the part of thousands of farmers' unions, consumer groups, NGOs, and faith-based and community organizations across the urban-rural and North-South divides of our planet. This remarkable "movement of movements" is widespread, highly diverse, refreshingly creative—and politically amorphous.

Many publications point to the hopeful initiatives in food production, processing, distribution, and consumption; and many analyses unpack and identify the structural impediments to a fair and sustainable food system. However, there has been little strategic reflection on just how to get from where we are—a broad but fragmented collection of hopeful alternatives—to where we need to be: *the new norm*. Unfortunately, social, environmental, and economic visions of what a good food system should look like are rarely accompanied by a clear political vision of how to get there. What can we do to roll back the corporate food regime and roll out the healthy, sustainable, and equitable transformation of the world's food systems?

In *Food Movements Unite!* food movement leaders from around the world answer the perennial political question: What is to be done?

The answers—from the multiple perspectives of community food security activists, farm and labor leaders, feminist thinkers, and prominent analysts—lay out strategies for convergence among the diverse actors and organizations in the global food movement. Authors take the corporate food regime head-on, arguing persuasively for specific changes to the way our food is produced, processed, distributed, and consumed. They also explain how these changes may come about politically.

In Part One, "Farmers, Sustainability, and Food Sovereignty," the producers of over half the world's food—family farmers and farmworkers—speak their minds in strong, clear, and radical terms. Paul Nicholson of the Basque Farmers' Union (EHNE), and João Pedro Stédile and Horácio Martins de Carvalho from the Brazilian Landless Workers Movement (MST) throw open the debate by describing the emergence and evolution of food sovereignty as a political platform to roll back the neoliberal assault of our food and farming systems. Basing their arguments on the rich experiences of agrarian struggle in the Basque Country, Brazil, and internationally, these leaders call for *alliances of transformative action* and *new structural policies* for our food systems. George Naylor of the National Family Farm Coalition provides a trenchant analysis of the corporate food regime from deep within North America. With a farmer's clarity, Naylor explains the opportunities and constraints on sustainable food production with the "Naylor Curve" and links historic US farm struggles for parity with today's international call for food sovereignty. Given that about 70% of the production and processing of food in Africa is done by women, Tabara Ndiaye and Mariamé Ouattara from West Africa explain that women's leadership is essential to "true food autonomy." They call for supporting an improved status for women within their communities, countries, and regions. These women play a central role in the grassroots campaign "We are the Solution" in Africa, a refreshing alternative to the corporate-driven "Alliance for a Green Revolution in Africa."

From the field, rural development experts from Africa, Latin America, and the Caribbean bring decades of experience working with farmers to the task of merging farmer-driven agroecological innovations with the political movements for food sovereignty.

John Wilson, a longtime sustainable-farming advocate from Zimbabwe, describes the spread of sustainable farming practices among NGOs and peasant farmer groups in East Africa. The gradual realization that farmer-led organizations will need to play a leading role in the transformation of peasant food production challenges NGOs to change their roles from purveyors of technical projects to supporters of political processes. The need to support farmers' political leadership is echoed in the final chapter of the section, written by the team from Groundswell, a new rural-development collective working in Haiti, Ecuador, Burkina Faso, and Ghana. Groundswell asks how NGOs can best support the movement for food sovereignty on the ground, among the farmers struggling to forge sustainable farming systems. A shift from project and donor driven strategies to farmer and movement-driven approaches is strongly advised.

Part Two, "Consumers, Labor, and Food Justice," shifts the focus to the Global North where farmers (who make up less than 2% of the population) are joined by food workers and consumers in a struggle to change the food system. Food justice, dismantling racism, plus healthy, local, and fair food strategies emerge as powerful progressive forces driving change. Activist-writer Raj Patel digs into the radical roots of the food justice movement in a retrospective on the Black Panther Party's free school breakfast programs. The Party's practical and political approach to neighborhood food security was part of a wider vision for social change. Josh Viertel of Slow Food USA calls on us to be active citizens in order to create a food system that is "good, clean, and fair" for everyone, not just those who can afford it. The structural barriers to healthy, equitable food systems are unpacked in detail by Brahm Ahmadi who, using the city of Oakland, California, as a case study, explains how "food deserts" are a reflection of the historical economic and political destruction visited on low-income communities of color. Ahmadi addresses the constructed divides of race and class, and calls for supporting the leadership of the underserved communities most affected by the injustices in the food system.

Lucas Benítez and José Oliva take on the issue of labor in the US food system. Indeed, one wonders how the food movement can even begin to think about transforming the food system without an understanding of labor's role in the present system and in the growing food movement. Farmworker and food-worker strategies for justice address actions and alliances with churches, universities, and other movements addressing labor rights and workers' food security throughout the food value chain.

Ken Meter argues that local food systems can and are playing a role in economic recovery in the US when consumers find ways to reinvest their food dollars locally. Case studies of local food businesses that build rural–urban links are followed by policy recommendations to strengthen local food economies. Meter's analysis shares the emphasis on the local power of the food dollar with Xavier Montagut's observations from the Catalán region on the Iberian peninsula. Montagut describes how Catalán and Spanish groups are keeping the power of the food dollar local—for both producers and consumers—through a radical system of fair trade. Unlike corporate-driven certification schemes that attempt to mainstream fair trade into the very corporate retail chains sucking food dollars out of the community, fair trade in Catalunya is a local-international strategy to keep the food dollar in the community by linking consumer sovereignty with producer sovereignty.

In Part Three, the final section, "Development, Climate, and Rights," draws from international and transnational movements for agricultural reform, climate justice, the women's movement, and the right to food. The right to food as a platform to transform our food systems is explained by Olivier De Schutter, who emphasizes the role of agroecology and social movements from within the vantage of the office of the Right to Food at the United Nations. Hans Herren and Anglea Hilmi of the Millennium Institute probe the potential of the pathbreaking—but corporately maligned—International Agricultural Assessment for Science, Knowledge and Technology for Development (IAASTD) to set a new agenda in which "business as usual is not an option." Nora McKeon elaborates on the struggle of social movements to access the agenda-setting power of the UN through the newly reconfigured Committee on Food Security. Linking climate justice to food sovereignty, Brian Tokar explores the natural convergence of these genuinely grassroots international movements—and calls for even greater North-South solidarity and alliance-building.

Taking the discussion to the realm of one of the world's most powerful social movements, Miriam Nobre shares the World March of Women's journey to food sovereignty and elucidates the ways that this has helped to shape both the women's movement and the food movement. Finally, Rosalinda Guillén—a farmworker and feminist—closes the section with a challenge for food activists to transform their own movements in order to transform the food system and ultimately, to transform themselves.

Bringing all of these authors together to address the question of "what is to be done" regarding our food movement has been an invigorating process and a sometimes daunting challenge. Food movement leaders are busy people with many urgent commitments. Many do not have much time to write. Fortunately, they are also visionaries with an insatiable thirst for justice. The threads of convergence running through the words of this diverse array of activists and practitioners, thinkers and doers from around the world weave a rich and dazzling tapestry of transformation. It is our hope that this will inspire readers to reach beyond their immediate concerns to view the food movement holistically, and to act accordingly by involving themselves more deeply in the processes that we all depend upon for our daily bread.

PART ONE:

FARMERS,
SUSTAINABILITY
AND
FOOD SOVEREIGNTY

Chapter One

FOOD SOVEREIGNTY: ALLIANCES AND TRANSFORMATION

By Paul Nicholson, Basque Farmers' Union, EHNE
International Coordinating Committee, La Via Campesina

This chapter is based on interviews with Paul Nicholson by Iñaki Bárcena Hinojal, Director, Department of Political Science and Adminstration at the University of the Basque Country, Euskadi, and by Eric Holt-Giménez, Executive Director of Food First.

THE CRUX OF ALL CONCERNS in the world today is food and the environment. Not only what we eat, but who is producing the food and how it is produced. Since one-sixth of the people in the world are starving, these concerns are growing.

The first indication of the string of global crises we have seen in the last few years (food, fuel, finance, and climate) was seen in the food system. These crises are related and systemic. They are forcing us to formulate new models for society, and agriculture is central to this necessarily extensive process of transformation. Eight years ago, we saw the emergence of the antiglobalization movement. This movement has filtered down to the local level, addressing public goods, water, and food—who produces, who controls, and the role that corporations play in the food system. The corporate race to monopolize land, seeds, water, and food stocks on the planet is not accidental. It is, in fact, the main issue. This is why food movements around the world are on the rise. There are an endless number of mobilizations worldwide that are quite diverse, yet they all speak about the same thing. Sometimes they don't even have similar demands, but they are aiming for the same goal. Across the industrial North and the Global South, local-level networks are becoming stronger as we create new ways to advance food alternatives and confront the global power of corporations.

Challenges

I am a dairy farmer and was in charge of the Basque Country Farmers' Union (Euskal Herriko Nekazarien Elkartasuna; EHNE) in Vizcaya when Spain entered the European Union. Agriculture was a fundamental part of these negotiations, and I was an observer in the process and participated in the debates. As an organization, we quickly realized what impact neoliberal policies were going to have on agriculture and food in the Basque Country. Understanding this—before people even talked about globalization—helped us comprehend that the decisions that affect the conditions and quality of life of our farmers are not made locally. The centers of political decision are anonymous and farther and farther away. We realized that we couldn't defend our rights in the Basque government, or Madrid, or Brussels, without a global vision. So the first secretariat of La Via Campesina was formed here, at EHNE's headquarters in Vizcaya, in 1994.

Since the battle in Seattle in 1999, social movements have been working to create alliances and strengthen networks, but there have been some difficulties in this historic process. Clearly, the World Social Forum is at a standstill over the need not only to be a reflection of people's power but also a political space from which to struggle. We'll see where we're headed, but I want to strengthen what exists in the thousands upon thousands of efforts taking place at the local level. Our challenge is this: How can we be the base of resistance and create a movement that actually presents alternatives to the present neoliberal system?

I believe that although neoliberal globalization has conquered the world economy, every time a minor resistance asserts itself it is a major event, because new movements are being formed. For example, the World March of Women is a very important movement. La Via Campesina, a movement of peasants, is also significant. Moreover, all of the alliances being created around climate change will be fundamental to growing as a movement.

We have to create common objectives—like food sovereignty—as a global alternative to the neoliberal model. Food sovereignty is not only about eating locally, eating well, and being able to afford to eat; it's also a food policy alternative to neoliberal policies because, if we want to change the well-being of family farmers, a simple band-aid will not work. We need to transform society. The whole debate over food, the environment, and property becomes a community question, and we have to consider this in order to form alliances. Alliances are something very complicated, but we have advanced greatly in the last 10 years. Right now we are at a bit of a standstill, so we have an opportunity to figure out how to bring our collective strengths together.

The Food Sovereignty Movement

Food sovereignty is a vision for changing society and, from a broad social and community perspective, an alternative to the neoliberal policies. It is the right of citizens to determine food and agriculture policies, and to decide what and how to produce and who produces. It is the right to public resources such as water, land, and seeds. Food sovereignty calls for policies based on solidarity among citizens and between consumers and producers. It calls on us to regulate markets because it is impossible to maintain agrarian policies based on market liberalization. Food sovereignty ensures socially sustainable, ecologically produced food that provides work for people everywhere.

This means that food sovereignty is much greater than food security. It's the political right to control policies and public goods, and to define what we eat from a social perspective, not just an individual one. And within the framework of neoliberal politics, this is clearly not going to happen. The theory and practice of comparative advantage has resulted in the massive destruction of the rural world, because it reduces everything to the criterion of competition without any structural consideration of social or labor rights. At the same time, it generates environmental costs that are then socialized—i.e., paid for by society, not by the companies that generate them. We have to show that neoliberal policies are the causes of the poverty, exclusion, and misery that exist in the world. And although we know that neoliberal policies have failed in this day and age, they continue to drive models of production that are absolutely destructive. The response we can give to such policies is food sovereignty, because it brings together movements from the industrial North, the Global South, and urban and rural areas. Food sovereignty as a human right is an integral demand from social movements around the world.

Alliances for Transformative Action

It is difficult to talk about alliances in the abstract. We can speak first of alliances around a general understanding, a common analysis, and common objectives—like layers of an onion. There are broad alliances, more specific ones, and there are alliances between organizations. These last are the most important ones. We need to make not only broad and specific alliances, but alliances for transformative action. We need to advance toward a common understanding that facilitates organizations such as La Via Campesina, the World March of Women, and Friends of the Earth that continue to assume transformative roles and are capable of action. For example, the World March of Women holds actions with a common message in thousands of places around the world.[1]

At La Via Campesina, we understand that the struggle around food is tied to the whole question of rights and will be one of the most important concepts for organizations and social movements around the world. The transnational corporations (TNCs) today

[1] More than 38,000 women have joined the World March of Women's actions since they began on International Women's Day on March 8, 2010. The WMW has developed four action areas: women's economic autonomy; common goods and public services; violence against women; and peace and demilitarization. See http://www.globalmarch.org/news/131010.php.

intervene directly in all our institutions and determine policies in favor of their own interests. They are our main enemy in this movement. The strategies against TNCs will be very diverse. We are speaking of not only financial capital, but also of the TNCs' involvement in the food chain—especially those that monopolize our food and determine the intellectual property of seeds. There lies the fundamental axis that we need to change.

Rural-urban alliances

First of all, the Food and Agriculture Organization (FAO) of the UN reports that 70% of the people in the world who are starving are farmers; 70% of these farmers are women. So the battle in most of the rural world is about food. It is a direct confrontation between the industrial corporate model and a peasant model of food production. Another crucial fact to understand is that most of the food eaten in the world is produced in the area in which it is eaten. There are different statistics, but around 60% of the world's food is produced and consumed in the same region. Finally, peasant agriculture feeds the planet, and we have the capacity to feed the entire world population both today and in the future.

It is industrial agriculture that creates hunger. In the most productive agricultural regions, where soy and corn are cultivated, we find the worst hunger. This example repeats itself everywhere. How we produce food, who produces it, and how we solve the food problem are not separate questions but are linked to one another. Today there is enough food for everyone in the world, but its distribution is unjust. What needs to be done is to localize the production near urban zones and break this corporate monopoly. We are convinced, at La Via Campesina, that agroecological farming not only cools down the planet (because it produces fewer greenhouse gasses than industrial farming), but also feeds most of the world. The problem is we are not reaping the profits from this. In Spain, the Coordinator of Farm and Ranching Organizations (COAG) did a study with consumer associations in which they monitored 40 products (from the various categories of fruits, vegetables, legumes, meat, and milk) on a monthly basis, comparing what farmers receive for their products and what consumers pay for food. This difference is enormous—as high as 1,000%! Moreover, this includes products such as onions

Sustainable Peasant and Family-Farm Agriculture Can Feed the World

La Via Campesina

The 2008 world food-price crisis, and more recent price hikes in 2010, have focused attention on the ability of the world food system to "feed the world." In La Via Campesina, the global alliance of peasant and family-farm organizations, we believe that agroecological food production by small farmers is the agricultural model best suited to meeting future food needs.

If we can agree that small-farm agroecological systems are more productive, conserve soils, restore the lost productivity of degraded systems, and are more resilient to climate change; then the key question is not whether we should, but how we can, promote a transition to such systems. Because agroecological systems require the mobilization of farmer ingenuity, the methods that work best are those in which farmers themselves become the protagonists in recovering, developing, and sharing methods. This can only happen inside farmer and peasant organizations, through farmer-to-farmer and community-based methods, farmer training schools, etc.

However, farmer organizations are swimming against the tide when we cannot count on effective public policies. Such policies must include genuine agrarian reform to put farm land into the hands of peasants and family farmers, end open and hidden subsidies to industrial farming methods, including chemical inputs and GMOs, reverse the free trade policies that make farming unprofitable, and make an overall shift from polices that are hostile to small farmers and their organizations to ones that support our own efforts to innovate and develop agroecological farming methods and share them horizontally. The time has come to act, to build true food sovereignty in each country based on agroecological farming by peasants and family farmers in control of our own destinies.

Full Article at: http://www.foodmovementsunite.com/addenda/via-campesina.html

and potatoes that don't have any further processing or handling. This means that what the TNCs are doing is monopolizing the profits of foods that they know perfectly well we need to eat.

Since there are similarities between the urban and rural movements, there is the possibility of creating another relationship between producers and consumers, between urbanities and farmers. They should seek an exchange of products (as is already happening in various forms around the world). Because farmers do not want to integrate themselves in the production chain controlled by the TNCs, they are looking to create short circuits and direct relationships with urban consumers. They want to make sure that their products reach consumers in the most efficient manner, with fair prices for farmers as well as consumers. The debate over price is central to the whole struggle against the neoliberal model.

We also need policies to deintensify production and facilitate a transition to agroecology—not organic or ecological certification schemes, but agroecological models that maintain employment in the countryside. For this we need new political instruments that regulate markets, establish fair prices, and regulate imports and exports. We are radically opposed to all direct and indirect subsidies for exports, but it is also our right to regulate imports. If the objective is to ensure a vibrant rural world and produce food, we need instruments for each task. This means that the policies must be more locally based, with new food-transportation models. For La Via Campesina, the struggle against large transportation infrastructure projects in Europe is a part of this effort, because such projects restrict our local productive capacity.

Political alliances to create our own alternative economies, processes, cultures, and environments

We are moving forward and opening many common spaces and alliances that did not exist ten years ago. But we are still very divided, each of us in our own niche. What can be done? We aren't just talking about eating locally or eating well. We are talking about much more: constructing alternative livelihoods based on local needs. Consumer organizations can't ignore the need for political transformation. This is not just an opportunity to ensure good local food. It is the beginning of a transformation of social relations between producers and consumers, of getting control, and of taking the initiative in what we do. Across Europe, there are many local experiences in which we are constructing our alternatives in different locales. But the big policies are done by central governments and that is where we have a big problem. The TNCs have a lot of power and influence that we have to weaken and eliminate. In all aspects, in all the so-called democratic institutions, we have to question the role of Europe and the US globally. The corporations are occupying places of decision in international institutions (e.g., the World Trade Organization, the World Bank) and in the European Union. The delegitimating of TNCs is a first-degree political challenge.

We also have to see how Europe fares with its financial crisis. We can already see that the global correlation of forces is going to change, and Europe will be weakened as an international institution. I think social movements have an opportunity here. It's a shame to see how the crisis makes us more reactionary or conservative regarding our privileges, but these privileges will no longer be sustainable, and deep social transformations are coming.

Over the last 30 years, we have seen a great distance divide us from political parties. In the last two years, I've seen a reconsideration of these earlier strategies; the formal political space is becoming of more interest. In the short term, the fact that French farm leader José Bové has a seat in the European Parliament is a good opportunity, and we will collaborate with him to take our struggle to the institutions. The social struggles in different countries have generated different political spaces. The experience of Evo Morales in Bolivia, and other peasant leaders with him, leads everyone to reflect on how to intervene in institutional political spaces. There are cases such as Nepal, Bolivia, and six to eight other countries in the world that are also references for political change. If there was a rejection of political intervention through political parties up to three years ago, today more and more we are seeing real possibilities for intervention, though not in Europe, where this is still very difficult. The formation of political parties is not a priority.

I have very little faith that the "corporate democracies" represent what we want today. I believe that we have to speak of a participatory democracy from below, from the social movements. I think that the creation of alternatives will be from the local level. The struggle has to link everyone together in the decision-making process.

This political fight is a struggle for knowledge, for culture, and for different ways of producing and consuming, all of which are part of us. For us, this is the struggle to rescue our own knowledge, not only of our own foods but also of our seeds. The biodiversity of seeds is impressive. Each region has its own seeds adapted to its climate and its agricultural and cultural needs. The rescue of this knowledge is another principal axis. The struggle to localize the production and the processing of foods is another principal axis, and that foods are under the control of the producers and the consumers. And how do we create this politically? I believe that it comes from the local level, from resistance and civil disobedience. We have to go construct our own realities now. We cannot wait for them to arrive from above. Transformation comes from the power of a process from below. The alliances that we create from below, from reality, come from our own proposals and actions. Ten years ago there was a struggle against the economy, and now we are struggling to construct our own alternative economies, processes, cultures, and environments.

Food Sovereignty to Solve the Food, Economic, and Environmental Crisis

Food sovereignty is fundamental. In 2010 the world's nations had three summits: the first was in Rome with the FAO, about the food crisis; the second was the World Trade Organization; the third was about climate change. At all three, the proposals of the corporations and the governments were the same: more technology, more trade, and more liberalization. But in none of these three was there a consensus. This presents an opportunity for food sovereignty.

We see that the food, economic, and climate crises are consequences of the neoliberal production and economic model. Food sovereignty is the alternative, from the perspective of the people and the social movements. This is the alternative that, moreover, proposes possibilities to solve problems related to climate change. The model of food sovereignty permits local production and productive agroecology models that can solve the climate problem and permit us to feed the world at the same time. It is a model of peasant and family-run agriculture that produces food principally for its surrounding populations. It doesn't produce agrofuels that exacerbate the problem.

In Spain unemployment is enormous, over 20% (Woolls 2010). In the Basque country, it is at 8.8% (EiTB News 2010). In a congress of ACE Vizcaya, a local organization, it was determined that food sovereignty could create an estimated 55,000 jobs. According to the farmer agricultural model, this produces food for a local market. We have proposed that 5% of the population should become farmers. In actuality, it is 1% of the population—almost at the point of disappearing! Our proposal is that through food sovereignty and agroecology, there will be 55,000 new farmers in this province. We understand this process clearly. For us, food sovereignty means creation, bringing in new farmers, both men and women. For this we need policies that promote the participation of young people. This requires a land bank that helps them get established. We are negotiating with the local institutions because it is essential to use the land to strengthen this decision. We have to provide land for agriculture and stop infrastructure policies (e.g., high-speed trains, highways, airports) that are wasting the most productive land. The new agriculturists need to be in the most productive regions. The whole debate about land and agrarian reform is important.

There has to be vision that is much more global. The present narrow vision is overshadowing the little memory we hold. When the food crisis happened, apparently this was the only worry that society had. It occupied the entire mass media that stated, "X amount of billions of dollars will be needed to solve the problem of hunger." Next appeared climate change, and the same people said, "This is going to cost . . ." But in reality it hasn't cost anything because everything has been forgotten. Now the financial and economic crises have the spotlight and the policies to resolve hunger have to wait, because the most important thing is to solve the financial crisis of capitalism. But it's this lack of memory, or mass-media domination, that controls the crisis itself. The main worries of the planet are about hunger and climate change. It's not the financial crisis, but all of the resources invested in solving the speculative model, creating another speculative model over the land, common property, and food. This is the next speculative bomb that will explode.

Works Cited:

EiTB News. 2010. "Unemployment rate falls slightly in the Basque Country" *EiTB.com News*, October 19, 2010. Accessed March 13, 2011. http://www.eitb.com/news/detail/524123/unemployment-rate-falls-slightly-the-basque-country/.

Woolls, Daniel. 2010. "Spain's unemployment rate passes 20%." *USA Today*, April 30, 2010. Accessed March 13, 2011. http://www.usatoday.com/money/world/2010-04-30-spain-unemployment_N.htm.

Chapter Two

PEOPLE NEED FOOD SOVEREIGNTY

By João Pedro Stédile and Horácio Martins de Carvalho[1]

João Pedro Stédile is an economist and member of the coordinating body of the Brazilian Landless Worker's Movement (MST) and *La Via Campesina,* Brazil. *Horácio Martins de Carvalho* is an agronomist, social scientist and consultant for *La Via Campesina* who looks at global food systems and Brazilian agriculture.

THROUGHOUT HUMAN HISTORY hunger has been associated with diverse phenomena such as poor food-production techniques, disputes over, and loss of fertile lands, natural disasters, and wars. During the 10th century these factors ceased to be the main causes. However, hunger and malnutrition affect more people today than at any other time in human history. Why?

The explanation is found in Josué de Castro's seminal thesis: "Hunger and malnutrition are not a natural occurrence but instead are a result of social and production relationships that humans establish amongst one another" (Castro 1951).

In fact, the hunger that in 2009 affected over a billion people worldwide has its root causes in monopoly control over production and distribution, in the income differences between people, and in the unequal distribution of profits from food production. Never before in human history has food been concentrated in a single production matrix as it is today. Fewer than 50 companies control most of the world's seed production, agricultural inputs, and food distribution worldwide.

---

[1] *We are grateful for the contribution of the following La Via Campesina researchers: Peter Rosset, Francisca Rodriguez, Pamela Caro, Irene Leon, Paul Nicholson, and Eric Holt-Giménez.*

The right to food for all people, independent of social condition, color, origin, sex, or age is no longer a human right in the international capitalist world. Today food access is restricted by the laws of profits and accumulation. People only have access to food if they have money to buy it. Because wealth is highly concentrated around the world—particularly in the countries of the Global South—the poor majority suffers from lack of food.

Food Sovereignty: Concepts and Trajectory

Over the last few decades there has been a positive evolution in the terms and concepts used to analyze hunger and malnutrition. For most of the 20th century the issue was treated as a problem resulting from natural phenomena. It was Josué de Castro's work *Geografia da Fome* (*The Geography of Hunger*), translated into more than 40 languages, which consolidated the concept that hunger is a social problem that is the result of the way in which society is organized and distributes its food. His theoretical contributions were so important that in the 1950s the United Nations granted him a post as the first secretary general of the FAO.

Later, in the 1990s, progress was made toward the concept of food security. This concept was formulated by governments and the FAO, so that in terms of human rights, all people have a guaranteed right to food. And it is the duty of governments to ensure this right. In this way all people would have the guaranteed security from governments that the necessary food for survival would be supplied.

This was an important step, because it is based on a public policy of obligation of all governments to resolve the problems of hunger affecting their populations. However—as the current food crisis demonstrates—it is insufficient.

More recently, the new concept of food sovereignty has emerged, introduced by La Via Campesina in 1996 at the World Food Summit (WFS) called by the FAO in Rome. The official debate around food security reaffirmed it as "the right of everyone to have access to healthy and nutritious food, consistent with the right to adequate food and the fundamental right to be free from hunger." However, farmers' organizations, and particularly the women delegates present at the parallel forum to the conference, were critical of the terms used in discussion with the governments. In line with neoliberalism and the WTO, the definition of food security sought to ensure food access through the liberalization of trade in food, opening opportunities for transnational corporations, the chemical industry, and fast food, among others.

The peasant organizations challenged the concept of food security with food sovereignty. They started from a prior principle in which "food is not a commodity, it is a human right" (Montecinos 2010) and the conviction that the production and distribution of food is a question of people's survival and therefore a question of public and national sovereignty. Sovereignty means that in addition to having access to food, the people, the populations of every country, have the right to produce. And it is this that will guarantee they have sovereignty over their lives. Control of production is fundamental for populations in order to have guaranteed access to their own food for the entire year. And so they can guarantee that this food is appropriate for the environment in which they live, for their nutritional needs and their food habits.

From there it evolved into a concept of food sovereignty whose meaning asserts that every community, every city, every region, every nation has the right and the duty to produce its own food. Despite the natural difficulties there may be, in whatever part of our planet, people can survive and live in dignity. The accumulated scientific knowledge needed to face and guarantee production and sufficient goods for this already exists.

And if production and distribution of food is part of the sovereignty of a people, it is nonnegotiable and cannot be made dependent on political will or the practices of governments of other countries. As José Martí warned, in the beginning of the 20th century, "A people who cannot produce their own food are an enslaved people. Enslaved and dependent on another nation to provide the conditions for survival!"

This concept breaks with agricultural markets imposed by transnational corporations and neoliberal governments and with the policies of the WTO and FAO. These policies have dismantled the policies of some nationalist and populist governments that sought to protect family farms through taxes on cheap imported foods, granting price bands and maintaining the power of public buyers (Montecinos 2010).

Food sovereignty counters the hegemony of neoliberalism by strengthening the vision of economic democracy in the world. This was affirmed at the world food sovereignty conference in Mali in 2007, in the Nyéléni Declaration (2007):

> Food sovereignty is the right of peoples to healthy and culturally appropriate food produced through ecologically sound and sustainable methods, and their right to define their own food and agriculture systems.

This situates those who produce, distribute, and consume food at the heart of food systems and policies, above the demands of markets and corporations. It defends their interests and takes into account future generations. It offers us a strategy for resisting and dismantling the corporate food regime, and for food production systems—agricultural, livestock, and fisheries—to be managed by local farmers and producers. Food sovereignty gives priority to local economies and local and national markets, and grants the power to farmers and family farming, fishers, and traditional pastoralists. It positions food production, distribution, and consumption on a sustainable, social, economic, and environmental base. Food sovereignty promotes transparent trade, a living wage for all peoples, and affirms the right of consumers to control their own food and nutrition. It guarantees the right of access and management of our lands, water, seed, cattle, and biodiversity, putting it in the hands of those who produce food. Food sovereignty implies new social relations free from oppression and inequality between men and women, peoples, racial groups, social classes, and generations.

Since food sovereignty is a concept that is built from popular sovereignty, it is absolutely incompatible with any strategy that tries to place private interests above those of the people. With neoliberal globalization, government control of production, processing, and distribution of basic foodstuffs has been replaced with overproduction and free trade policies supported by large national and transnational agribusiness corporations. This has increased risks to food security, as the logic of supply submits to the interests of corporations that control national and international markets. This undermines the autonomy that regions have always had in the production of food, and puts sovereign food production, practiced by local peasants and small and medium-sized farmers at risk.

The social organizations that introduced the term *food sovereignty* emphasize that it is more than just a concept. It is a principle and an ethical lifestyle that does not correlate with an academic definition but arises from a collective, participatory process that is popular and progressive, and whose essence is constantly enriched through various agrarian debates and political discussions in the peasant organizations. La Via Campesina, founded in 1992, and its Latin American member Coordinadora Latinoamericana de Organizaciones del Campo (CLOC), founded in 1994, were originally the main organizations advancing this principle. The Food Sovereignty conference in Mali also consolidated a broad alliance with other social movements of fisherfolk, pastoralists, urban and rural women, consumers, environmentalists, nutritionists, researchers, scientists, public health movements, and progressive governments that continue to collectively build new understandings of food sovereignty.

The various collectively articulated documents and declarations of food sovereignty elaborate upon the right of peoples to define their own agricultural and food policies, including: the protection of the environment; the regulation of production in fisheries and internal agricultural trade for sustainable development; the protection of local and national markets against imports and limits on market *dumping* of social and economic products. It embodies the right to decide *how* to organize *what* is produced and how to organize food distribution and consumption in relation to the needs of communities, in sufficient quantities and qualities, prioritizing local products and local variety.

For Francisca Rodriguez of Anamuri, a Chilean peasant organization, "Food sovereignty is linked to the other struggles we are involved in: the struggle for land, to keep our culture and our ways of life, to make our work and contributions as women visible and valued" (Rodriguez 2010).

More recently, at the World People's Conference on Climate Change and the Rights of Mother Earth (2010) in Cochabamba, food sovereignty was ratified as referring to

> The right of peoples to control their own seeds, land, water and food production, ensuring, through local production, autonomous (participatory, community and shared) and culturally appropriate, consistent and complementary with Mother Earth, the peoples' access to sufficient, varied and nutritious food as well as deepening the production of each nation and people.

This reaffirmed new visions and concepts based on the "Good Living," or *Sumak Kawsay*, a concept born from ancient Andean heritage and deeply woven into the fabric of Andean popular grassroots organizations. It is consistent with the rights of peoples to control their own territories, their natural resources, and their social reproduction, and of integration between ethnicities and peoples in accord with common interests—not just as determined by trade and profit. There is also influence on the concept from a feminine perspective of the world of fertility and social reproduction of humanity in egalitarian and just conditions.

The declarations and agreements constructed in the forums, seminars, and national and international conferences worldwide on *food sovereignty*, with participation from the majority of civil society, women's peasant movements, and some government sectors, have not yet, unfortunately, resonated in practice in the transformation of public policies of most governments and international bodies.

The Brazilian Case

Brazilian society still suffers from serious structural problems, so much so that in its production and social organization, it is still unable to ensure food sovereignty to its people. For many years the statistics showed that more than 50 million Brazilians suffered from hunger every day.

The causes of this have been thoroughly analyzed in many studies, tests, and investigations in academia, newspapers, and public institutions. We can synthesize these by saying that the unjust and inequitable structure of wealth produced and concentrated over the past 500 years of capitalism has resulted in an extremely unequal society, in which 5% of the population controls most of the wealth. The distribution of annual income is unjust, with most of the profit going toward capital and much less for the workers. The distribution of assets—especially land—is unjust, with only 1% of property owners controlling 46% of all land (Stédile 2010); a consortia of large capitalist companies, based in cities and with a focus on other financial activities, controls 170 million ha of land. Additionally, according to the most conservative estimates, foreign capital has already appropriated more than 40 million ha of Brazilian land.

The factors impeding food sovereignty in Brazil are many, starting with the model of production and agroindustrialization the country introduced decades ago. In this sense, the position on food sovereignty in the Declaration of Brasilia (2008) is clear:

> We maintain that hunger and poverty are not products of chance, but rather produced by a model that violates the rights to a decent life for people and peoples, increases the subordination of women and exploits workers and their social, economic and cultural contributions to society. Despite the

worldwide evidence of the disastrous effects of the neoliberal model, the international system, the governments and multinational corporations insist on submitting the planet to a development model that exhausts the possibilities of life, turning people into mere agents of production, without faces and without histories. Economic liberalization, as the only path to development, is directly proportional to the growth of poverty and hunger in the region; the failure to exercise food sovereignty seriously undermines the sovereignty of states.

The concentration of land ownership in Brazil has reached 0.57, according to the Gini index and the UN's 2006 Agricultural Census (higher than its concentration in the 1920s). Nearly 178 million ha of pristine land—one-third of which is now in a state of degradation—have seen the marked expansion of genetically modified crop plantations, accompanied by the displacement of the peasantry due to the social and physical pressure of large monoculture estates of soy, sugarcane, corn, and eucalyptus. This situation has led to peasant demands for land and local markets—as reflected in the Declaration of Nyéléni. Food sovereignty reflects demands for nutritious and culturally appropriate food. Since the drastic reduction of the government's role in the agricultural sector in the 1990s (and despite the presence of some public agencies and programs such as Companhia Nacional de Abastecimento (CONAB)and Programa Nacional de Fortalecimento da Agricultura Familiar (PRONAF), Brazil is home to the largely unregulated expansion of corporate control in agriculture.

The privatization of seeds through the imposition of genetically modified crops deepens corporate control over agricultural production, in violation of the right to conserve the collective heritage of seeds, which is at the core of people's food sovereignty.

The agribusiness mode of production, based on large-scale monocultures, imposes the permanent and ever-increasing use of pesticides. Brazil turned into a major global consumer of agrotoxins, consuming one billion liters of pesticides in the

Land Sovereignty

Jun Borras and Jennifer Franco

In the context of converging food, financial, and energy crises and increasing global land grabbing, land sovereignty is an alternative analytical framework that represents the defensive struggle of the rural poor. Land sovereignty is the right of impoverished peoples to effective access to, control over, and use of land to live on as both a resource and territory. Thus, as a pillar of food sovereignty, land sovereignty is the right to land.

The metanarrative of neoliberal capitalist governance is naturalizing global land grabbing, as both necessary to secure food and energy supplies, and as an opportunity for rural development. In response, an international "Code of Conduct" (CoC) among stakeholders is proposed as a "win-win" economic development strategy for both wealthy investors and the rural poor. Proposed by the World Bank, FAO, IFAD, and UNCTAD, the CoC proposal is not inherently pro-poor and diverts attention from root problems with the economic development model. The CoC proposal distracts from rural poor people's interests and land rights.

A shift from conventional land reform, namely land-tenure security, to land sovereignty is critical to achieving effective control by the rural poor over land where they live and work. Land sovereignty requires starting from existing land-based social relations to work toward clear pro-poor land reforms. Land sovereignty politicizes and historicizes land issues, providing a counternarrative to neoliberal governance.

A people's counter-enclosure campaign is one in which the rural poor resist the inequitable status quo in land ownership and control, particularly land grabbing. A people's enclosure campaign is one in which the rural poor proactively assert their political control over lands against potential and actual threats of elite enclosure. A land sovereignty movement is necessarily a "people's" (counter)enclosure movement: rural people will both resist inequalities (i.e., land grabs) and assert political control over remaining lands.

Full Article at: http://www.foodmovementsunite.com/addenda/borras-franco.html

2009/10 growing season (an average of six liters per person and/or 150 liters per ha). Agribusiness is feeding us contaminated food, destroying our biodiversity, affecting our water quality and the air we breathe, and driving climate change. This is the real tragedy.

This tragic reality was partially addressed in 2003, when the president-elect of the nation confronted the public with the problem of hunger in Brazil. The government engaged the popular sectors in the re-creation of the Board for Food and Nutritional Security (Conselho Nacional de Segurança Alimentar e Nutricional; CONSEA) and by holding various conferences. They launched the Zero Hunger (Fome Zero) campaign, which consists of prioritizing a group of programs and actions of diverse ministries to confront food insecurity in Brazil.

Within the Zero Hunger campaign the government implemented Bolsa Familia, or Family Food Basket, the Food Purchase Program (Programa de Aquisição de Alimentos; PAA), and the deployment of a network of food security teams and equipment, such as soup kitchens, community kitchens, and food banks throughout the entire country. They also implemented a National School Food Program (Programa Nacional de Alimentação Escolar; PNAE), which includes the direct purchase of products from family farmers, and PRONAF (the National Program for the Strengthening of Family Agriculture).

Although the number of Brazilians going hungry has been reduced because of Zero Hunger, the causes of the problem were not addressed, and therefore hunger persists. Though they do not suffer severe hunger, some 60 million Brazilians still do not get enough food.

The current food security policies are important, even if they are insufficient in terms of attacking the root of the problem. The compensatory government programs such as the Family Food Basket contribute to the food security of millions of poor people. However, this tends to be confused with food sovereignty policies, when in fact, due to their cyclical nature, they are really emergency actions to minimize food insecurity due to extreme poverty.

The dominant logic of capital in Brazil is to assign the population's food supply to the commercial interests of large national and international corporations in the food sector. This means a subordination of national sovereignty to profit and to private, oligopoly interests. However, this contradicts the historical concept of national sovereignty, which is the fundamental framework for a sovereign nation. However bright and colorful the practice implemented may be, private interests should not govern sovereign public claims, even in terms of food.

The conclusions of the third National Conference on Food and Nutritional Security (CNSAN) in Fortaleza, Brazil, in 2007, emphatically reaffirmed that the objective of food and nutritional security implies an approach to socioeconomic development that questions the present hegemonic model in Brazil that drives inequality, poverty, and hunger, and that negatively impacts our health and environment (CNSAN 2007).

Structural Policies to Attain Food Sovereignty

The basis of building food sovereignty in Brazil, within the more general construction of national sovereignty, requires structural reforms in rural areas and reforms to the country's current model of agricultural production. We would like to close by reaffirming the platform for structural reforms adopted at the CLOC's 5th Continental Congress held in Quito, Ecuador, in 2010:

- A broad and massive agrarian reform to democratize the ownership and use of land, with the consequence of ensuring four million working families access who want to produce in agriculture. For this it is necessary to expropriate the largest estates and especially the lands of foreign capital and of nonagricultural companies, banks, etc.
- Change the current model of production and dominant agricultural technology to different concept of healthy food production, with a basis on agroecology, ecoagriculture,

organic agriculture, and others to ensure production and abundant supply in all locations, regional and at the national level.

- Limit the size of property and land ownership; ensure the principle interests of the whole society in relation to natural, water, and biodiversity resources.
- Restructure the role of the state to organize the process of food sovereignty, ensuring the production and distribution in all of the country's regions.
- Direct control of government on foreign trade (import/export) and food, interest, and exchange rates.
- Implement a comprehensive program of small- and medium-sized agroindustries in all municipalities in the country, in a cooperative manner.
- Ensure buffer stocks of healthy food by the government to guarantee access to the entire population.
- Develop a new economic model, based on wide distribution of income, secure employment, and income for the entire population, with universal education and implementation of industry oriented for the national market.
- The knowledge and liberty to exchange and create better seeds is a fundamental component of food sovereignty, because diversity assures food abundance, with a base in adequate and varied nutrition, and the development of culinary forms which are culturally appropriate and desired. Seeds are the beginning and the end of the cycle of life for peasants, they are collective creations that reflect a history of a people and its women, who have been its creators, primary protectors, and "perfectionalists." Their disappearance leads to the disappearance of cultures of the countryside and of communities. Because they are not appropriated, they shall maintain their character as a collective asset.
- Prevent the use and promotion of genetically modified seeds. They represent the privatization of life, of the possibility of

free reproduction and, above all, they represent the destruction of all biodiversity, since they do not reproduce without the contamination of all other seeds. In addition to this is the doubts and lack of research of their effects on animal and human health.

- The right of peoples and of all Brazilian people to consume according to cultural, ethical, religious, aesthetic, quality food, which implies healthy, affordable, and culturally appropriate food is a "sine qua non" to achieve genuine food sovereignty.

Works Cited

Castro, Josué de 1951. *Geopolítica da fome*. Rio de Janeiro: Casa do Estudante do Brasil.

CLOC (Coordinadora Latinoamericana de Organizaciones del Campo). 2010. *V Congreso de la Coordinadora Latinoamericana de Organizaciones del Campo, Quito, Ecuador, 2010*. Accessed February 22, 2011. http://www.cloc-viacampesina.net/es/pronunciamientos/2010/404-clocomunicacon.

CNSAN (Conferência Nacional de Segurança Alimentar e Nutricional). 2007. *Declaração Final da A III Conferência Nacional de Segurança Alimentar e Nutricional, realizada no Centro de Convenções do município de Fortaleza (CE).*

International Planning Committee for Food Sovereignty. 2009. Declaration of Nyéléni. 2007. Accessed February 24, 2011. http://www.foodsovereignty.org/Portals/0/documentisito/Resources/Archive/Forum/2007/2007-En-Declaration_of_Nyeleni.pdf.

La Via Campesina. 2007. "Sementes, patrimônio dos povos a serviço da humanidade." Documento Campanha em defesa das sementes da La Via Campesina.

Montecinos, Camila. 2010. Personal interview.

Nyeleni.org. 2007. "Declaration of Nyéléni." *World Forum on Food Sovereignty. Sélingué, Mali.* Accessed February 22, 2010. http://www.nyeleni.org/?lang=en.

Rodriguez, Francisca. 2010. "The Seeds of Sovereignty." *New Internationalist Magazine.* Accessed June 1, 2010. http://www.exacteditions.com/exact/browse/386/422/7513/3/3

Stédile, João Pedro. 2010. "A Republican and Democratic Reform is Necessary in Brazil." *Interview with the Landless Workers Movement* (Movimento dos Trabalhadores Rurais Sem Terra; MST). Accessed March 22, 2010. http://www.mstbrazil.org/?q=node/649.

World People's Conference on Climate Change and the Rights of Mother Earth Working Group 17: Agriculture and Food Sovereignty. 2010. "Final Conclusions." Accessed February 22, 2011. http://pwccc.wordpress.com/2010/04/29/final-conclusions-working-group-17-agriculture-and-food-sovereignty/.

Chapter Three

WITHOUT CLARITY ON PARITY ALL YOU GET IS CHARITY

George Naylor,
National Family Farm Coalition

"WE CAN HAVE ANY kind of agricultural system we want," was a statement made to many farmers across the United States by the late Merle Hansen, a family farmer from Nebraska, back in the days of the 20th century when he was a leader in the American Agriculture Movement and North American Farm Alliance. Another slogan was also often heard: "Without clarity on parity, all you get is charity." Merle returned home from World War II knowing that by the law of the land, farmers deserved and would receive parity, and he was familiar with his family's participation in decades of family-farm organizing and struggle to get those laws on the books.

Parity simply means equality—which is why it's consistently vilified by free-market agriculture economists and agribusiness publications. Since the goal of parity was erased by the passage of the Hope-Aiken Act in 1954, and many of the final nails in the coffin driven in family-farm agriculture by the Freedom to Farm Act of 1996, the "charity" came in the form of subsidy payments and government-sponsored "economic development" programs to employ farmers who sadly had to work off the farm or leave the farm altogether (Pratt 1996). Wendell Berry (2009) spells it out in an excellent article entitled "Inverting the Economic Order": "The farm population has now declined almost to non-existence because, since the middle of the last century, we have deliberately depressed farm income, while allowing production costs to rise, for the sake of 'cheap food,' and to favor agribusiness."

Despite the difficulties of farming today, whether you're working on a Corn Belt farm that's been in the family for almost a century, or starting a new organic CSA (community supported agriculture) for local consumers, we can be glad for the birth of a new food and farm movement that is both democratic and international in scope. Not only is it concerned with fairness for farmers, fishers, and workers, and with healthful food for consumers, but it recognizes we all must live within ecological limits and plan ahead to leave a beautiful and supportive ecosphere for future generations.

Recent history, with the industrialization and monopolization of food and farming, has offered plenty of environmental, social, and political horror stories, which motivate us all to "choose the agricultural system we want." Creating local markets where farmers and consumers get to know each other can counter our alienation and the dehumanizing corporate marketing schemes. But we must also make choices collectively on a local, national, and even international level.

As William Greider (1993) explains in *Who Will Tell the People*, we now toil and consume in a global economy under the tyranny of the World Trade Organization (WTO), whose dominant ideology betrays the real substance of democracy (as illustrated by the 2010 Citizens United decision of the Supreme Court). He says, "This has produced a daunting paradox: Restoring the domestic political order will require a new version of internationalism." In other words, can we ignore the fact that the can of pumpkin in the local grocery store is now produced in China, or that the fresh green beans were picked in Guatemala, where nearly half of the children are malnourished? Can we believe that farmers in the US benefit when corn and soybeans are shipped to Third World countries to feed chickens and pigs in giant factory farms owned by a few multinational corporations?

I believe our new unified food movement will lead the way in this restoration and recognize that people in other countries are looking for the very same kind of democratic choices we are.

The international peasant and farmworker movement La Via Campesina has a name for it: food sovereignty. Every country should be allowed to have food sovereignty, so that each country can democratically develop its own policy to choose the kind of agriculture its people want, and ensure its citizens' food security and political sovereignty. With food sovereignty, a country can make sure that its food production is ecological and economically fair, and provides economic opportunity in rural as well as urban areas, so as not to be subject to the whims of global markets, global corporations, or the use of food as a weapon by more powerful countries. After all, why should a country and its people lose its sovereignty and security by having to import food from the "winners" who are willing to depopulate rural communities and pollute and destroy their farmland with pesticides and unwise production at any cost?

I'd like to offer some insight I've gained about why and how we might make collective decisions to choose the agricultural system we want. I'm lucky; I can draw from my mom and dad's family history of farming in Iowa for over 100 years, including my own 35 years, and from farm leaders like Merle Hansen and his family, who worked to pass and retain the New Deal parity farm programs, and from many friends working in the trenches producing food for CSAs and farmers markets. I'd like to tell the consumers out there: there are some great farmers you ought to get to know, with whom you'll hit it right off, in the National Family Farm Coalition (NFFC) and La Via Campesina.

For too long, multinational corporations have foisted on us free-market ideology that says we should ignore the importance of federal policy and international trade policy or, worse yet, that such policy should not even exist. Many times we fall for the political spin from corporate economists or politicians pretending to be on our side, leaving their lobbyists and economic hitmen to conquer our food system. Their power has grown as world hunger continues, and ecosystems like the US prairies and South American rain forests and savannas fall victim to corporate bulldozers and genetically engineered monocrops. Without clarity on parity, all you get is charity.

Free-market solutions, or what US politicians have referred to since 1953 as "market-oriented" policy, won't serve the interests of family farmers or consumers who need healthy food. "Supply and demand" actually delivers more of the farming and food nightmares mentioned above.

Let's first recognize that the power of multinational corporations and even farmers' co-ops has grown so much that they can now generate new production of commodities and livestock just by making the internal decisions, because it furthers their desire for vertical integration or strategic growth. The Spanish organization GRAIN has documented "land grabs" in Africa and South America by other countries such as China and Saudi Arabia, often in alliance with multinational corporate partners.

One unbelievable example is that of a US farmer's co-op, CHS, that created a joint venture in Brazil. According to Corn & Soybean Digest (2007), "CHS Inc., America's largest farmer cooperative based in Inver Grove Heights, MN, announces it has teamed with partners in Brazil and Japan to buy 385 sq. mi. of Brazilian farmland to produce soybeans, cotton and sugarcane." Not only would this be in competition with its farmer members in the US, but the farming was going to be done without farmers! Six hundred employees were to be hired, which surely would break new ground for a model that would be anathema to the original intent of the co-op.

Also in the US, family-farm livestock farmers have long protested the practice by the multinational meat packers of expanding livestock production at the same time family farmers are getting disastrous prices on the free market. The companies do this through direct ownership of new feedlots or sweetheart production contracts with already giant feeding companies. Brazilian meatpacker JBS, which claims to be the "largest animal protein company in the world," made 14 acquisitions around the world between 2007 and 2010 (Korby and Russo 2011). JBS bought Swift from ConAgra and purchased America's largest feedlot, which can handle 850,000 heads at a time—47% of the cattle it raises as a producer! Some are calling this practice "captive supply." These corporations can, in effect, increase production at any time to depress the open-market prices that farmers get, and then use that price to even ratchet down prices received under contract. Data indicate that less than 6% of hogs are priced on a free-market basis (Ellis 2010). Some of the organizations working against captive supplies and demanding enforcement of new rules under the Grain Inspection, Packers and Stockyards Administration (GIPSA) are the Western Organization of Resource Councils, R-CALF USA, the National Family Farm Coalition, and Food and Water Watch.

Fruit, vegetable, and diary producers face similar monopolized marketing situations today, since New Deal solutions have been systematically abandoned. These programs offered help to fruit and vegetable farmers through marketing agreements, ensuring produce from all sizes of farms were inspected fairly and had equal access to the market. Dairy farmers were also ensured regionalized parity prices. Such programs could be revitalized for today's problems.

The 900-pound gorilla in the room is commodity production (approximately 266 million acres versus 12 million, for all fruits and vegetables). What is usually meant by "commodities" is nonperishable grains and oilseeds that can be stored or shipped at any time of the year. Grains are both "food grains" like wheat or rice, and "feed grains," of which corn is king. Feed grains also include sorghum, barley, oats, and even wheat, when cheap enough. The benchmark international prices for these commodities are set at futures exchanges like the Chicago Mercantile Exchange while the "basis," the difference between Chicago and the local price, is influenced more by local conditions and transportation costs to various destinations. These are also often referred to as "program crops" because US farm programs traditionally address the economic issues of these crops most directly.

Commodities are usually considered "fungible"; that is, one bushel of corn is as good as another. If there is a difference, then federal grading standards are used to discount prices accordingly. Also, commodity prices are very interdependent because farmers in some regions can easily switch production between various alternatives. On the demand side, protein, carbohydrates, and oil found in various feed grains and oilseeds can be interchanged with little difficulty, especially for livestock feed.

I often run across the argument that, if we just had more competition, more grain buyers and processors, the commodity prices would likely be fair. Another argument is that farmers overproduce commodities but not fruits and vegetables because of government subsidies. I don't believe these are adequate arguments. The truth is that our country's history from very early times has included long periods of depressed prices (overproduction) and land degradation. In modern history, farm prices collapsed after World War I, before the existence of any farm program, driving farmers into poverty while the rest of society experienced the Roaring Twenties. This country's endowment of rich prairie soils always made the production of commodities an unwise temptation that could never be permanently remedied. Also, since commodities can be stored, they can be converted into cash any time of year, unlike fruits and vegetables, which are perishable. Fruits and vegetables have been made cheap at the farm level and industrialized with cheap energy, unwise pesticide use, cheap farm labor and subsidized irrigation. Simply creating subsidies for fruit and vegetable production could create an oversupply and even cheaper prices for new local producers. Besides, US imports more than doubled between 1996 and 2006, to reach $15.4 billion (Krissoff and Wainio 2007).

Monopoly power undoubtedly gives corporations the ability to adjust their production decisions more rapidly, by shutting down plants, laying off workers, manipulating futures markets, or unduly influencing regulators. But I believe the most pressing issues, such as prices being too low or too high, or the detrimental use of the land, can be explained pure and simple by the basic workings of farmer behavior in pure competition—producing their "supply" in response to expected "demand." In other words, contrary to much of our economics training, "supply and demand" is not the holy grail that can solve all agricultural, resource, or labor problems.

The esteemed ecologist Edward O. Wilson has indicated that modern civilization's demands on our natural ecosystems and the resulting loss of biodiversity and ecosystem services threaten civilization itself. As Frances Moore Lappé has always insisted, enough food is produced in the world to provide everyone with an adequate diet, but excessive grain and oilseed production results in wasteful feedlot production of meat, milk, and eggs along with processed foods that feed (unhealthily) the richest members of the global economy. Wilson indicates that nations, in fact the global community, must begin to take inventory of natural resources, including biodiversity, and make conscious, collective decisions about economic use and conservation. In other words, international economic cooperation combined with food sovereignty is our only hope to eat well, live well, and live within our ecological limits. The following discussion will hopefully give us the understanding and tools to take our unified food movement forward.

The Naylor Curve

Michael Pollan (2006) facetiously refers to my thinking on agricultural policy and family-farmer behavior as "the Naylor Curve." This is a supply curve, a graph that aims to predict a family farmer's production response (Quantity) to changes in commodity prices (Price). Many economics textbooks simplistically use examples of agricultural supply curves sloping forward at 45 degrees and intersecting with a demand curve perpendicularly, thus leading to incorrect conclusions on market behavior and government policy needed to create a sustainable economy.

The textbook supply curve implies that a farmer will increase production in proportion to price increases and decrease production in proportion to price decreases. This may apply to a "firm" as used in nonagricultural textbook examples, where production occurs day by day, and a firm can rapidly adjust use of capacity and labor in response to price signals.

By contrast, a family farmer has a large, fixed investment, and basically only family members, who cannot be laid off, for labor. Since family farmers are price takers (i.e., the decisions of a single family farmer will have no impact on the prices all family farmers receive for corn, soybeans, wheat, etc.), the farmer's revenue depends solely on aggregate output—the more bushels, the larger the income. The real-world supply curve that most closely resembles the textbook example relates to a farmer who intends to maximize profit no matter what the cost to the environment, family, or community, and has no problem using hired labor. In this case, the curve would slope upward, but it could be very steep—indicating that such a farmer has already applied nearly all technological products with diminishing returns. Also, any major change in production might involve bulldozing woodland or plowing grassland that hasn't been in commodity production before, thus shifting the supply curve in discrete steps to the right.

In understanding more common family-farm production decisions, other considerations must be taken into account. Unlike many other sectors of the economy, where production transpires day by day, once the farmer plants the crops in the spring, no adjustment in production is likely to take place. For commodities (grains and oilseeds), farmers have a high fixed investment; much of the variable cost occurs early in the growing season so that harvest will most likely proceed (something to sell is better than nothing to sell).

Another consideration is that demand for food is very inelastic, which means that even when prices go lower, very little new demand is developed, and only very slowly. In other words, the farmer cannot count on low prices to automatically create the demand to correct the low-price situation (Ray, De La Torre Ugarte, and Tiller 2003).

Actually, individual supply curves that aim to reflect actual economic behavior need to take into account complex cultural considerations, a farmer's values, expectations, ability to assume risk, and government policy. Let's first look at farmer behavior in an economic framework of "market-oriented" farm policy, where farmers are supposed to make decisions based on market signals and expected prices. Commodity futures prices change moment by moment at the Chicago Mercantile Exchange and are monitored by farmers around the world as benchmark prices. Price volatility, exacerbated by speculation, currency devaluations like we are currently experiencing, and unpredictable international weather events make price expectation very imprecise.

Let's also remember that in this case we are talking about a family farmer who does have something of an environmental ethic, values time spent with family, and would like to leave a farm as a thriving family asset for future generations. In other words, maximizing profits is not everything. I think we can recognize the likely responses to changes in expected prices in terms of segments on the graph, so that the final curve obviously won't look like the textbook's forward-sloping supply curve of the firm.These segments reflect generalizations of various zones of commodity price expectation with some sense of stability (not often realized in a free-market situation), but you must also understand that the farmer's behavior might be different, depending on whether the farmer expects prices to be increasing or decreasing (known as hysteresis).

The following graph will be used only schematically, to illustrate my thinking, and could be actually more complex with more intense analysis.

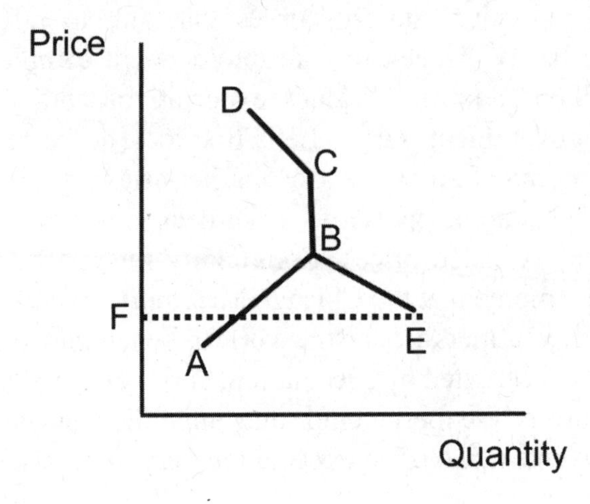

AB—limited-resource zone: For the beginning, limited-resource farmer trying to scrape by, better prices afford better inputs and implements to increase production. Years of depressed farm income mean that today there are relatively few commodity farmers in this category (many of the "new farmers" produce products other than commodities marketed through alternative channels). The following segments refer to farmers who have passed the stage of beginning, limited-resource farmers.

BC—comfort zone: At these price levels, the family farmer receives a comfortable income that rises when prices rise without increasing production. This can take place within a comfortable relationship to land, community, and neighbors if there is an expectation that prices will not fall so low as to threaten the family's standard of living or, even worse, force them into bankruptcy.

CD—biodiversity-restoration zone (BRZ): As prices rise above the comfort zone, a family farmer reaches an income level where he can actually decrease production and devote more time to leisure and/or use land for alternate conserving uses, including developing regional relationships to restore biodiversity. Once again, this will only happen if the farmer believes prices will not drop below this zone. The graph shows that if prices do start to decline, the farmer is likely to increase production, implying behavior fitting a backward-sloping supply curve. The current boom in commodity prices could put many family farmers in this bracket. Nevertheless, the expectation that these prices will stay secure in the long term is not commonly held, resulting in some farmers acting like the profit maximizer, using increased income to increase production through more destructive practices, such as using fungicides or leveling trees to bring the land into commodity production. As more and more farmers incrementally increase production, and multinational agribusinesses encourage production on other continents, the free-market result will, in fact, be the lower prices predicted.

BE—survival zone: As indicated above, as price expectations slide downward through the comfort zone, farmers may feel production increases are necessary to protect their families' standard of living. Prices below the comfort zone beget fears of further declining prices and a substandard living or possible bankruptcy, and further motivate family farmers to increase production. They contribute more hours and increase chemical and mechanical inputs, thus putting more pressure on the land (resource degradation). The curve could step drastically to the right if a farmer is tempted to bulldoze trees or plow up other land never previously used for commodity production. Gains from the recent price boom would quickly evaporate. Most existing farmers could find themselves in this zone.

In past years, even government "deficiency payments" (a.k.a. subsidies), which were never adjusted for inflation, didn't bring family farmers into the comfort zone. Fear of declining prices and income remain, and once again the supply curve is sloping backward, i.e., declining prices indicate increasing production. If the vast majority of farmers follow suit, prices may decline even further, leading to a poverty-resource degradation cycle, which will be discussed later.

BF—collapse of farming operation: At this price, the farmer simply quits farming and production equals zero, except for the likely outcome that the land will be rented by another farmer who tries to survive by spreading fixed costs over more acres, or implementing new technologies that might increase soil erosion, destroy soil health and biodiversity, possibly even bringing marginal land into production.

What the Naylor Curve actually shows is that under "market-oriented" farm policy, farmers will either base their behavior on two motivations: greed or fear. There are government regulations in other areas of our economy that at least nominally limit behavior to keep greed from disrupting orderly marketing or preserving our environment. That issue must be faced honestly through resource policy, but outcomes due to economic fear must also be dealt with through the economic framework of the Farm Bill or international commodity agreements that can replace the disastrous "market-oriented" straightjacket of the WTO.

The political reality is that giant agribusiness has lobbied for and achieved a "market-oriented" policy since 1953, institutionalizing uncertainty and forcing farmers to be willing adapters of new, expensive, and often destructive technology, motivated to increase production at all costs to family, community, and environment. This cynical mindset was expressed by the right-wing Secretary of Agriculture Ezra Taft Benson, during the Eisenhower administration of the 1950s. He spoke out against the policies of the New Deal, promoting the idea of "getting the government out of agriculture" and coining the phrase "freedom to farm." One of his more infamous and cynical statements was, "Farmers need the spur of insecurity" (Benson 1960; Wallace 1943).

A good farm program, as implemented with New Deal principles, would aim to establish minimum prices in the "comfort zone" so that farmers didn't have to fear ending up in the "survival zone." From 1941 to 1953, the price floor for storable commodities was set at 90% of parity (in this case, the parity price for each commodity has been indexed for inflation to a price at an earlier time that seemed in balance during the "parity base years"). During that time, farm prices overall averaged 100% of parity, and because the purchasers were paying the price, the government actually made a small profit (rather than costing the Treasury billions of dollars, as in later years when farm programs substituted government payments to make up for low prices). In abundant years, surplus production was stored to maintain the price floor in a government food security reserve, the Ever Normal Granary, to be brought onto the market to prevent speculation and hoarding at, say, a level of 120% of parity. We could implement such a program today to keep farmers in this comfort zone, the parity zone.

Breaking the parity guarantee in 1953 created another very troublesome consequence. Since corn and soybean protein are the major ingredients in manufactured livestock feed, their overly low prices have led to increased livestock production and exceedingly low livestock prices. Diversified family farms that produced most of their own feed and wisely recycled nutrients were hurt financially, thus shifting livestock production from family farms to industrial feedlots (CAFOs) that bought and shipped the cheap feed all over the country and even the world. When the diversified family farms gave up raising livestock, there was no longer any need for hay, pasture, or small grains that were integral to soil conserving and nonpolluting crop rotations. Farmers then often plowed up these conserving crops to raise more corn and soybeans! This led to soil erosion, water pollution, and lower commodity prices which led to lower meat, milk, and egg prices, and so on. Scenes of topsoil sliding down a hill across a road have been witnessed firsthand by dairy farmers like John Kinsman, president of Family Farm Defenders, in Wisconsin. Development economists refer to this type of vicious cycle as the "poverty-resource degradation cycle," but it is often only recognized happening in Third World countries.

Looking to the Future

Actually, society could decide democratically that we need to restore our biodiversity and should live on a healthy diet that doesn't require current production levels or deplete so many natural resources. We could eat less but better-quality meat, milk, and eggs produced on many more sustainable farms. We have the option, in our quest to make the kind of agriculture we want, to increase the price of unhealthy processed foods made from cheap corn, soy, wheat, and dairy products. The price floor would be pushed higher along with increased supply management and incentives (financed by all taxpayers) into the biodiversity-restoration zone, even the biodiversity- and healthy-diet-restoration zone (BHDRZ).

It should be noted that the very first New Deal farm program, the Agricultural Adjustment Act, was declared unconstitutional by a business-oriented Supreme Court because the act focused on price guarantees and taxes on processors. The program was rewritten to pass constitutional muster by focusing on the need for conserving our natural resources. This should help us recognize the connection between economic justice and environmental stewardship today. One of the best methods of restoring biodiversity and achieving supply management would be to issue each parcel of land a quota of commodities to be marketed or consumed as livestock feed—each farm's fair share of the new national market. Let's say that in the BHDRZ, we democratically decide we need one-third fewer commodities at a much higher price. The farmer would work with a local production committee to figure out a plan to produce two-thirds of current levels while minimizing fertilizer, chemicals, and unwise tillage, and maximizing conservation and habitat restoration. This simple change in the structural framework of farming can erase the normal mindset of trying to maximize output motivated by fear or greed.

Since the average farmer today is nearly 60 years old and often doesn't own most of the land he or she farms (cash rents now rob farmers of income and independence), it shouldn't be hard to convince farmers and landowners that new, smaller farms with new environmentally minded farmers are in order. Giant feedlots could systematically be made obsolete. The billions of dollars saved by eliminating farm-subsidy programs could be used to compensate older landowners and CAFO owners for new agrarian reform. New, small-scale, local production would gradually be ushered in along with new local processing plants and markets (currently promoted by Food and Water Watch). The cheap-food policy highlighted by Wendell Berry would end up in the dustbin of history!

Intergenerational Dialogue and International Solidarity

So how are we going to prove Merle Hansen's assertion that "we can have any kind of agricultural system we want?" The current state of politics in this country is not encouraging, but being quiet and acquiescent is not an answer. One thing about Merle was that he could communicate with young farmers like me and young people wanting to be farmers. He would talk about the vision many of us had of becoming organic farmers. His experience and knowledge of the history of farmers organizing and struggling inspired us all. It was not hard to understand that common farmers of all generations and countries aren't all that much different, unless they believe the right wing's version of freedom, that we just want to be able to do whatever we want without recognizing our responsibilities to each other and our beautiful planet; that democratic government can't help us achieve what's needed for the common good.

Organized efforts are underway across the nation to pass the torch—inspiration, knowledge, land, and resources—to new, young, and beginning farmers. For example, the Farmer-Veteran Coalition aims to "assist the farming community by developing a new generation of farmers and to help our returning veterans find viable careers and means to heal on America's farms."

The Federation of Southern Cooperatives has led the fight with African American farmers to stem the loss of black farmland.

Ben Burkett, president of the NFFC and staff director of the Mississippi Association of Cooperatives, joined Ralph Paige and a delegation of farmers and civil rights groups on December 8, 2010, at the signing by President Obama of the legislation for a $1 billion settlement to resolve the decade-long discrimination claims against the US Department of Agriculture. Similar lawsuits have been settled that deal with issues relating to Latino and Native American farmers. These lawsuits are a step toward stemming land loss and land grabs, and will hopefully shift the direction toward all famers and growers having access to the land and capital that they deserve.

The current world "food crisis" punctuates the suffering created when bad agricultural policy drives people from the land, and rural communities become dependent on uncaring labor markets in giant metropolises. Giant agribusinesses, like Monsanto and DuPont, churn out bogus arguments justifying the use of genetically engineered crops that kill off the biodiversity of precious natural ecosystems and rural human landscapes. This time around, if we don't get parity, we may not get even charity.

Our vision of a beautiful, restored ecosystem can surely inspire action. It was once the law of the land that farmers deserved and would receive parity and economic justice without making food scarce or plundering our land. International commodity agreements were recognized as essential to post-World War II peace and prosperity by respected economists and leaders like John Maynard Keynes, Henry A. Wallace, and Benjamin Graham (known as the Dean of Wall Street). What else could be more important than dialoguing and working with our fellow farmers, peasants, farmworkers in La Via Campesina, and our fellow citizens around the world to achieve food sovereignty, to achieve real democracy as members of a thriving biosphere?

Works Cited

Benson, Ezra T. 1960. *Freedom to Farm*. New York: Doubleday & Company.

Berry, Wendell. 2009. "Inverting the Economic Order." *The Progressive*. September.

Corn & Soybean Digest. 2007. "CHS Buys Brazil Farmland." Accessed March 17, 2011. http://cornandsoybeandigest.com/chs-buys-brazil-farmland.

Ellis, Shane. 2010. USDA fourth quarter Iowa, Southern Minnesota Pig Report. #LM_HG204.

GRAIN (Genetic Resources Action International). 2010. "World Bank Report on Land Grabbing: Beyond the Smoke and Mirrors." September. Accessed March 23, 2011. http://www.grain.org/articles/?id=70.

Greider, William. 1993. *Who will tell the People: The Betrayal of American Democracy*. New York: Simon & Schuster.

Russo, Camila and Boris Korby. 2011. *Sara Lee Takeover Bid Leaving JBS Bonds Out of Market Rally: Brazil Credit.* Accessed May 12, 2011. http://www.bloomberg.com/news/2011-01-05/sara-lee-takeover-bid-leaving-jbs-bonds-out-of-market-rally-brazil-credit.html.

Krissoff, Barry, and John Wainio. 2007. *U.S. Fruit and Vegetable Imports Outpace Exports*. Accessed March 23, 2011. http://www.highbeam.com/doc/1P3-1277613521.html.

Pollan, Michael. 2006. *The Omnivore's Dilemma: A Natural History of Four Meals*. New York: Penguin.

Pratt, William C. 1996. "The Farmers Union, McCarthyism, and the Demise of the Agrarian Left." *The Historian*. 58: 329–42.

Ray, Daryll E., D. G. De La Torre Ugarte and Kelly J. Tiller. 2003. *Rethinking U.S. Agricultural Policy: Changing Course to Secure Farmer Livelihoods Worldwide*. Agricultural Policy Analysis Center. University of Tennessee. Accessed March 17, 2011. http://www.agpolicy.org/blueprint/APACReport8-20-03WITHCOVER.pdf.

Wallace, Henry A. 1943. *The Century of the Common Man*. New York: Reynal & Hitchcock.

Chapter Four

RURAL WOMEN CREATE THRIVING FOOD SYSTEMS IN WEST AFRICA

By Tabara Ndiaye and Mariamé Ouattara, The New Field Foundation, Edited by Deanna Drake Seeba

Introduction

THE FOLLOWING analysis regarding factors of hunger and strategies for food security in West Africa is taken from interviews with New Field Foundation Program Consultants Tabara Ndiaye of Senegal and Mariamé Ouattara of Burkina Faso. While their examples sometimes reference other countries, their experiences are rooted in their communities of origin, specifically Casamance, in southern Senegal, and Banfora, in western Burkina Faso, on the border with Cote d'Ivoire. They also draw on their ongoing work and relationships with women farmers' associations and networks in different parts of West Africa.

Given that about 70% of the production and processing of food in Africa is done by women, the opportunities and challenges lived by rural women in West Africa are inexorably linked with the state of food security. As Ndiaye and Ouattara explain, rural women's integral role in supplying food means that true food autonomy is not possible without their leadership and improved status within their communities, countries, and regions.

Rural Culture and Traditional Knowledge

While specific practices and norms differ between West African communities, certain characteristics of life for rural women are common across the region. One of these is the heavy burden of work performed by women and the lack of familial or community recognition for this work.

"I was so shocked at the age of 10, when I went to live in my aunt's village, to see women in rural Burkina suffering so much from morning to evening, and not receiving anything like the recognition they deserve," explains Ouattara.

"The daily lot of a rural woman in Casamance is work," echoes Ndiaye. "During the rainy season, if you visit a village, you won't see a single woman who isn't in the fields. They leave at five in the morning and they return at seven in the evening, and then they have to prepare dinner for the family, if they do not have a daughter to do this for them."

In Casamance, women traditionally hold a very important position as providers of food in the community in general, and in the family in particular. Agriculture, at least smallholder agriculture, relies fundamentally on women farmers because they are responsible for growing the food to feed their families. Specifically, this means providing the daily dish of rice and some vegetables for family consumption. Rice fields are therefore traditionally the domain of women farmers, as are grain stores, which are managed by women. If a woman's harvest does not sufficiently feed her family, traditionally she suffers the disapproval of her community. So women consider it a matter of honor to produce enough rice to feed their families.

Similarly in Burkina Faso, women are producers and providers of food for their families. As such, they experience a greater sense of responsibility and urgency when there is nothing to eat. In a situation of poverty, even if men are supposed to provide money daily, when the man has nothing, he still expects his wife to find something for the family to eat. With this responsibility, women play a central role in food production, preservation, and consumption.

In West Africa, women are necessarily implicated and concerned with food systems because it is they who strive to achieve food security on a daily basis. In all families, a young girl is taught from an early age to set something aside for the next day. This means that when she is given grain to make the day's meal, she puts a portion aside to make sure there will be something for the following day. Any woman or mother must be capable of managing resources so that everyone has at least a little each day, instead of everything being eaten at once, with nothing for the future. This is the basic principle of food security and it is already at work in rural communities.

The commitment to food security goes far beyond the family. In almost every country in West Africa, there are well-organized movements that are seeking to achieve food autonomy and security. One major force in the region is the *Réseau des organisations paysannes et de producteurs de l'Afrique de l'Ouest* (ROPPA – Network of Farmers' and Agricultural Producers' Organizations in West Africa), which brings together hundreds of peasant organizations in each country to ensure food security. It is not that West African governments are ignoring food security. These states have defined food security policies. They have established institutions to ensure food security. But they have not paid particular attention to the centrality and contribution of rural women to food security. Rather they tend to ignore the role of rural women. For example, Burkina Faso's national food security policy is not based on in-depth research about rural women's contribution, yet women have a lot of knowledge to share. Be it in Senegal, Burkina Faso, Mali, Niger, or elsewhere in West Africa, women have long established strategies for food security that have evolved over generations. National and regional movements for food security could also benefit from consulting rural women. However, the low status of rural women means that this is not considered important.

When the state does attempt to engage rural women in food security issues, it tends to do so in the form of instruction. In Casamance, agricultural extension workers and technicians who are employees of the Senegalese government often arrive in a village with new seeds, chemicals, and fertilizers and tell women farmers what they must do, without asking what they think, or without explaining clearly the reasons for using these new inputs and methods. Many rural women will accept what is being given to them, because it comes from the government and because they are told to by educated males. At the same time, many are interested in retaining ownership of their traditional ways of farming. So they choose to keep their traditional seeds, practices, and knowledge intact, and continue to sow and conserve their own seeds even when government services and agricultural research agencies provide new varieties of seeds and fertilizers at little or no cost.

Although the majority of rural women are still practicing non-mechanized agriculture with traditional hand tools, which can be slow and very tiring, there is the sense that these practices are time tested by their parents and grandparents and that over the long run, they are healthier and more sustainable. With a small garden plot close to their home, combined with rice production and other staple crops such as millet and sorghum, they are able to feed their families, even if the harvest is not very large. Additionally, as farmers become aware of the effects of climate change, they are developing and applying techniques that both preserve the environment and intensify production. Agro-forestry, for example, allows farmers to integrate animal rearing, agriculture and forestry that increase their yields while protecting their natural resources.

Some organizations are working to help facilitate this process of preserving traditional seeds. The *Association Sénégalaise des Producteurs de Semences Paysannes* (ASPSP – The Senegalese Association of Producers of Peasant Seeds) is an association based in Senegal whose members are peasant farmers who work to keep inventories of traditional seeds that exist in a given location. ASPSP partners with local peasant organizations to train them on how to preserve and store their seeds for the next season. By using pilot farms, they do practical demonstrations for rural organizations to instruct them in how to carry out agro-ecol practices. The women who attend these trainings then repl these methods on their own plots and instruct other women in their organization. Through seed fairs, which are internationally attended, ASPSP encourages farmers to exchange seeds and discuss traditional agricultural methods.

In many countries in West Africa, support for traditional seeds is compounded by a fear of genetically modified organisms (GMOs). In general, rural women are afraid of GMOs. They say GMOs cause malformation in babies, and an increase in disease, including cancer, and they say they have noticed that in the long term their yields have fallen with the use of GMO seeds. The publicity for GMO seeds is viewed with suspicion and actually serves to catalyze a mobilization for the fight against them. Most rural women's groups will categorically refuse GMOs as a viable alternative. In addition, women farmers often state that the use of chemical pesticides and fertilizers changes the taste of the food they grow, and reduces the length of time that the food will stay fresh, thus reducing the value of their harvest, because it rots easily.

Factors of Famine

Given the centrality of rural women to food production, any barriers they face become barriers to food security. There are many factors that have created and continue to perpetuate famine and food insecurity in West Africa. These challenges come from all levels, from the state through to the community level. From the start of a rural woman's life, there exists an unequal social relationship which dictates that every woman must be accountable to a man. This is to say that each woman is under the domination of a male relative from the time she is born until she dies, be it her husband, her father, her uncle or her son. Some women are dominated by all four.

Indicative of this dynamic is the cultural attitude that women do not need to attend school. For a long time, states did nothing to reverse this trend. Fortunately, for the past ten years or so, states have begun instituting policies to ensure that girl children are also educated. However, there is still work to be done at the family level to convince parents that girls need to attend school. It is said that 'if you educate a girl, you educate a whole nation.' We have seen that a woman who attends school changes things around her more quickly than one who has not had the same educational opportunities. Although both an educated and an uneducated woman might have the same potential, the lack of education is enough of a setback to keep the unschooled woman from succeeding as quickly.

Many rural women today are working to become literate, either in French or in a local language, so they can access information, for example on the Internet. They understand that this is an important resource for information and for building collaborations with other groups in Africa and around the world. They know that mobilizing is the way to address issues at both the local and regional levels. Just because a woman is illiterate, it does not mean she is any less capable of analyzing and strategizing than the holder of a doctorate. For this reason, men are wary of allowing women too much independence. As we say, 'if the women wake up, countries will tremble.'

There is, therefore, much resistance within families and communities to the involvement of women. For example, if a man does not allow his wife to go to her group's meeting, he may do so out of fear. A typical strategy for dealing with this is for the woman to highlight her role as the wife and mother in the house. She will show that she does her work at home well, that she provides enough food, and will even undertake some small-scale commerce in order to bring additional resources into the home. Women who participate in associations will also sometimes bring home small quantities of money from group activities, or foodstuffs from communal production. With these successes, the wife is then able to negotiate her participation in meetings again. This type of subtle strategy demonstrates that rural women do indeed know their rights, but given current cultural and political biases against them, they must employ caution in order to gradually exercise them.

Another form of prejudice is at the policy level. Often development strategies at the state level are not designed to address the real needs of rural people, particularly rural women. Often these policies are designed to meet international imperatives. For example, when there is a crisis, such as the food crisis, the energy crisis, the political or economic crises, states will immediately define policies to address them and not include the realities that their populations face.

The results of such oversight are the many symptoms of food insecurity. Farms and production systems are not sufficiently well organized; access to production tools is complicated and becoming ever more so; and the distribution channels for food are not sufficient to cover a country. For example, in Burkina Faso, there is a surplus of cereals in the west of the country, but because there is no means to transport the crops, there is a shortage in the center and north of the country. In this way, whole regions fall victim to policies that do not incorporate their needs.

Another example of state intervention to meet foreign imperatives comes from Senegal. For many years, Senegal received aid in the form of rice and cereals from other countries, particularly Asian countries. Rather than developing local agriculture to support the population, the state accepted donations from other countries to stave off famine. The cereals and rice that were brought in, however, were of varieties that the people in Casamance had never seen before, and so they were not easily able to use them. Although there are exceptions, even today the norm is to 'help' local rural populations without consulting them. Such intervention can be extremely costly and have little or no tangible effect on lessening a crisis.

The most basic and prevalent example of this dynamic of exclusion is that, by and large, rural women in West Africa do not own the land they work on. They are loaned land by a male relative or other community member. This quickly becomes a factor of the food crisis when the titled owner of the land decides there is more profit in selling to a foreign state or corporation than allowing women to continue agricultural production. More and more, fields that have been worked for generations are being sold off or taken over in what has become known as 'land grabbing.' Without a place to cultivate food, women cannot prevent famine at the family and community level and they are forced to leave their homes, often for the city, to try and make a living for their families through domestic labor, as sex workers, or laborers on others' fields.

Migration has had a huge impact on the ability of rural communities to produce enough food for the nation as a whole. During and after colonization, many people considered Casamance to be the bread basket of Senegal—that is to say, it was capable of feeding the entire country. But during the years of conflict in that region, there was much displacement from the villages across the border to The Gambia and Guinea Bissau, and to the towns and cities in other parts of Senegal. Suddenly there was no workforce to farm in Casamance. Famine quickly spread and the situation was very long lasting. As peace was established, people began returning to their villages but they found that most of the cultivatable land was now riddled with anti-personnel mines, which made it too dangerous to farm. Additionally, many young people who had fled to the city during the conflict did not return, which means that the strongest workers are no longer present to help with cultivation. This inability to return to previous levels of agricultural production is one of the main causes for ongoing hunger in Casamance today.

Women's Associations and Collective Action

To assert their agency and face these challenges, rural women across West Africa have begun to leverage their collective power through their groups and associations.

"The cultural specificity of Casamance is that each woman who is born in a village belongs to an association. Even when women from Casamance are outside their regions, they get organized into associations," explains Ndiaye.

Belonging to an association is common for women across West Africa. "Women need to become organized so that they are stronger," agrees Ouattara. "In their organizations, they learn more about their rights as women, along with strategies to achieve these rights for themselves."

Through these 'unions,' or associations, women play an important role in the food system, from production through to consumption. They produce, preserve, and process food so that it is available all year round. They select and prepare what their families eat. They organize to give each other advice, to gain access to technologies and resources, and to be in a better position to deal with cultural, sociological, economic, political, and legal obstacles to their well-being.

In order to find solutions to the difficulties they face—the lack of resources, poverty, lack of knowledge, lack of equipment, lack of financial resources—women have naturally grouped themselves into organizations across West Africa. There are small associations everywhere in Casamance, for example, and now the new trend is that those smaller groups are banding together into networks. This trend is proving to be very beneficial for the groups because the larger and stronger that organizations become, the more they are able to capitalize on major opportunities.

One example is the *Directoire National des Femmes en Elevage* (DINFEL – The National Organization of Women in Livestock Breeding), in Senegal, a network of more than 20,000 women members who are raising livestock. Established in 1996, the organization has 12 branches in different parts of Senegal that influence local policies and practices in relation to livestock and agriculture. In the capital, Dakar, its leaders advise the government on pastoral and livestock matters and have successfully influenced policy on land distribution, milk subsidy, and international markets.

Networks need not be based in state capitals to be effective, however. One important opportunity organizations are benefitting from is the current political process of decentralization in West Africa. As political power becomes less concentrated in the central government and moves out to the level of districts and villages, women are able to have a greater influence on community resources and management. Decision-makers are now their neighbors, their husbands, and members of their communities. Women's associations organize themselves and develop positions and strategies to approach these decision-makers. As a result of the decentralization process, all members of the community have become advocates for themselves in order to ensure that their needs are met by village councils. Women have also been seizing opportunities to carry out advocacy within the framework of decentralization. They support campaigns to elect preferred candidates and to monitor the actions of decision-making bodies within their communities. They also stand for election themselves, often making a public commitment to remain accountable to rural women and to prioritize their needs.

As women band together into larger organizations, their capacity to attract national and international partners also increases. For example, a large number of rural women's groups are now being supported by New Field Foundation through its grantmaking program, *Rural Women Creating Change*. New Field identifies larger organizations that are capable of receiving and managing international grants and that have track records of working with rural women's groups. In turn, this organization makes many smaller community grants to women's associations that do not yet have the same level of organizational infrastructure. By supporting these smaller associations with resources and technical services, the larger community grantmaking organization empowers them to meet their needs, plan for the future, and begin to improve the lives of their families and communities. Overall, rural women's groups know what is needed for their success and, with a community grant, they have the power to carry out their plans.

Increasingly, what women's associations and networks want is their own agricultural equipment for reducing their workload, particularly when it also brings economic benefit to the group as a whole. This includes motorized cultivators, drip irrigation systems, solar pumps, rice hulling machines, and tractors. When women's associations own and manage this equipment, it becomes evident that that they can quickly increase their acreage of production while reducing their labor, thus giving women time to engage in additional activities that benefit their communities.

Without title to land, however, it becomes complicated for women's groups to take advantage of their resources. More and more women are taking part in capacity-building programs to enable them to advocate for land ownership. They are not attempting to achieve individual ownership, but collective ownership on behalf of their associations and members. It is customary for women to work the land together for several years and produce a yield; this is a traditional practice and they do not consider it essential to seek title to the land. But when a rural women's association gets a community grant and learns how to improve its farming, they go to their local authorities and explain that they will not invest in the land until they hold the title to it.

It has been the experience of many groups that, when they improve the land (for example with irrigation), the man or family that has the rights to the land will come to claim it. This may mean that crops have to be cut down before the harvest. In order to mitigate this risk, community grants now incorporate an advocacy component so that associations are empowered to collectively own the plots they work. There is great progress being made in Casamance with this strategy, with a number of associations now having title to land, which is remarkable considering that not very long ago it was unimaginable from a cultural perspective for women to own land.

A common factor in all these successes is the availability of capacity-building opportunities for women's associations. In many West African countries, agricultural policies and programs have been geared, since independence, towards supporting men and young people, not women. As the primary producers of food, however, women are the more obvious target for such programs. Women's farming knowledge and expertise would usefully inform the design of agricultural programs whereby relevant agricultural information would be delivered directly to women farmers, without intermediaries. Instead, women have been consistently denied the opportunity to contribute their knowledge to a wider sphere, and to engage in the discussion of appropriate methodologies.

In analyzing the needs and successes of the groups New Field works with, we have come to realize that what they are lacking is not just financial and production resources but also support in terms of training and direct engagement with information.

"From my perspective, women still need capacity-building," says Ndiaye. "They need financial resources, but that alone is not enough. Although there are many women's associations, most of them are organizationally weak. Even if a group can mobilize resources among themselves, they often lack the knowledge and infrastructure to make the best use of these resources."

This is the kind of support New Field has offered through its community grantmaking program—to enable rural women to open and operate bank accounts; prepare proposals and budgets; track income and expenditure; produce reports; prioritize and manage activities according to plan. Just six years after the community grantmaking program started, stories of success are starting to emerge from the communities that have received them—an increase in diverse food production and income, more children attending school, and better health. Some women from within the communities have emerged as group leaders and are widely respected and consulted on issues affecting women and their communities, including agriculture. Their presence at meetings ensures that women's concerns are heard. For example, women's associations have effectively advocated local politicians to de-mine their fields, rather than focusing only on the areas where men have their agricultural activities. Elsewhere, rural women leaders are members of advisory committees to inform the implementation of international funding, often on a large scale.

This success should not be a surprise. The potential for women's leadership already exists within West African communities and families. Women all over Casamance will tell you, "We are leaders because we have leadership in our home and in our families. We exercise this leadership, for example, when we decide that our daughters will remain in school because we have the funds to pay tuition."

When rural women can show they have produced enough food, not just for their own family but for other members of their association and community, they hold a kind of power that effectively enables them to reduce the violence they suffer. When a husband sees that his wife is participating financially in the household and meets the family's needs, he often is less violent. For women, this is the meaning of peace. We think of peace as the end of a conflict, but if you ask a woman in Casamance, "What is peace to you?" she will tell you that peace is, first of all, in the home. When her family can eat properly, when her husband does not hit her, and when she can decide that her children will attend school because she can afford the textbooks, then she knows there is peace. These are common daily examples of women's leadership.

Community-wide Food Security

The advantage of working with women's groups is that often they are from the same locality or the same village. As agricultural production increases for the group, the nutritional needs of the community as a whole are met, not just those of individual families. As production increases, members of the groups become food secure, they begin to store food for the future, they give a portion to others, and they bring a portion to the market for sale. While this is often considered by agricultural scientists and development experts to be small scale, relevant only to the village or 'subsistence' level, we have seen how a large number of families and villages can become food secure, and how whole communities and even regions can achieve the same level of food security through this approach.

Word spreads quickly. Women's associations that have received community grants are now attracting women from other villages who have heard of their success. They arrive and say, "We've come to learn from you, how did you do it?" and the women answer, "We organized ourselves, we're a group." When the visiting women return home, they begin to organize themselves and follow the example of the groups that have already experienced success. Development partners in turn find it easier to work with rural women's groups that have already been formed and are functioning well.

There could be even greater impact with more community grants and capacity building to support women's associations that are organizing effectively. As they develop effective food systems that bring stability to their community, other work can also be done, such as peace-building, leadership training, advocacy, or financial and organizational management. This kind of training is not as effective, however, if there is no existing activity that federates women, such as agricultural production. In other words, additional projects have come in to capitalize on the strength of the women's federations. The women now have the time and the resources to be able to undertake other activities that help them to improve their lives, and the lives of their families and communities. Women leaders are now consulted by local authorities as part of decision making processes that affect whole communities.

Global Alliances for Food Sovereignty

The most effective method of achieving thriving food systems in West Africa is to support rural women to bring it about themselves. It is not a question of outside agencies or someone else doing it for them. Rural women have the resources in terms of local seeds and knowledge to create food security for their families, associations, communities, and countries. If we continue to keep all policy decisions and assistance coming from above, these traditional resources are at risk of being lost. If women's organizations and networks are given the space and the resources to develop local food systems that build on their existing knowledge and resources, widespread food security can be achieved in West Africa. But such change must come from the grassroots, from rural women and their networks. To foster their success, however, it is mandatory that policy decisions be taken in a participatory manner and that agricultural production activities are completely local.

Organizations that want to support communities to develop food security must give rural women and their organizations enough leeway to decide what they will produce and how they will produce it. They need to be trusted to make choices for themselves and their associations. This will result in communities that are no longer hungry. Although each community functions on a relatively small scale, if more and more build local food systems and attain a level of food autonomy, then many parts of West Africa will become food self-sufficient.

The question remains: what can the global movement for food sovereignty do to foster this progress? It starts with the recognition of rural women's role in food production and an acknowledgement of their place in the movement. Then organizations must consult these women and their organizations about their actions and priorities. Rural women represent a rich bank of information and can develop good strategies for food security and autonomy. They are capable of establishing linkages across groups and sharing information about techniques and experiences on a significant scale. They must be empowered to gain access through information management and technology so that their contributions can enter the international dialogue on food sovereignty.

A current example of this process can be found in the campaign for food sovereignty called "We Are the Solution: Celebrating Family Farming in Africa." The movement was conceived during the 2007 Nyéléni food sovereignty conference held in Bamako, Mali. At the gathering, farmer organizations expressed concern over the introduction of the new green revolution in Africa. In the face of the food crisis, initiatives such as the Alliance for a Green Revolution in Africa (AGRA), a multi-million dollar project launched by the Bill and Melinda Gates and Rockefeller Foundations, presented a new challenge to peasant smallholder farmers, especially women who are practicing traditional agriculture.

The Green Revolution approach can create dependence on high yield seeds, as well as costly chemical pesticides and fertilizers, and has been shown in many parts of the world to drive smallholder farmers off the land and into greater poverty. It can also include the conversion of large areas of land to grow crops that feed the global marketplace, whether plant materials for bio-fuels, bulk food commodities such as soy, rice, cocoa, coffee and palm-oil, as well as non-food products such as flowers and cotton. This industrial-style farming has the effect of draining resources away from smallholder farmers, threatening local biodiversity and diminishing regional food security.

In response to the challenge of a new Green Revolution in Africa, "We Are the Solution: Celebrating Family Farming in Africa" emerged as a grassroots, farmer-led campaign that aims to mobilize and support an Africa-wide network of 1,000 stakeholders through information sharing, active partnership and advocacy. An initial focus on Women and Biodiversity encourages and supports rural women's organizations and networks to assume leadership responsibilities as part of the Pan-African campaign. It emphasizes the recognition of the traditional knowledge of African women in agricultural production and supports them to build their capacity to share this knowledge and advocate for its wider recognition as a viable alternative to the Green Revolution.

The project will rely on rural women's organizations already engaged in the practice and promotion of ecological agriculture to enhance their expertise and their activities and share best practices on a larger scale.

"We Are the Solution!" strives to bring together many organizations and individuals to work together towards a common goal. The campaign is part of a broader movement in West Africa and around the world with a mission to achieve food sovereignty, protect the heritage of crops, fight industrial-style agriculture, ensure land rights for smallholder women farmers, promote the rights of producers and end hunger.

Collaboration and a united movement—at community, national, regional, and international levels—are vital for success. But real success is only likely to happen if the millions of smallholder farmers whose lives and livelihoods are at stake across the world, including West Africa, are integral to and help lead the process. This means active recognition of and respect for their tremendous creativity, energy, and capacity at community and national level. Funders and policy makers can take positive action by establishing genuine partnerships and investing in the grassroots actions, leadership and capacity of local farmers and their organizations—particularly those of rural women with the ability, knowledge and commitment to feed their communities, and get food into their local markets. There is potential for spectacular success on a scale never seen before.

Chapter Five

IRREPRESSABLY TOWARD FOOD SOVEREIGNTY

By John Wilson, Farmer/facilitator from Zimbabwe
co-founder of PELUM Association of East Africa.

TODAY IN EAST AND SOUTHERN Africa there is a new "putsch" by industrial agricultural proponents, but there is also a gathering resistance to this, which has grown steadily over the last 25 years. Today, we have the opportunity to unite practitioners, advocates, and consumers around a powerful food sovereignty movement that offers viable alternatives, and influences policies to benefit small farmers and their communities across the region, instead of benefiting corporations. Can we seize this moment?

I write this chapter as a personal essay, based on the credentials of having been a small part of the organized development and spread of sustainable agriculture as a practice in East and southern Africa. This essay is a reflection about where we have come from and where we are going, the story we have written and the story we *could* write. I say *we* as a rallying call primarily to people in the East and southern African regions, but also to people all over the world who want to join forces as part of a global food sovereignty movement. The emphasis in this essay is on the positive things we can do together. Though I do not dwell on it here, I am only too aware of the cut and thrust, and strength in financial resources of those who continue to push an approach to the land that is not sustainable and that contributes to greater inequity.

Sustainable Agriculture in East and Southern Africa: The Last 25 Years

In February 2007, La Via Campesina, the organization that originated the idea of food sovereignty, hosted a meeting at Nyéléni in Sélingué, Mali. Out of this came the Declaration of Nyéléni (2007), an evocative statement calling for a global food sovereignty movement. In November 2009, seven African umbrella organizations met in Addis Ababa, Ethiopia, to launch the Alliance for Food Sovereignty in Africa (AFSA). Both these events were momentous steps toward citizens across Africa regaining control of their food, or of keeping this control in some cases (Food First 2009). In 25 years' time, I believe we will see these events as key turning points in an unfolding story.

Twenty-five years ago, the dream across Africa was still for industrial agriculture to sweep the continent. This represented the ultraliberal orthodoxy of the day during what are often referred to as the Reagan-Thatcher years, that period of modern history when the free market became sacrosanct. In agriculture, it was simply a question of assisting small-scale farmers to follow in the tracks of large-scale commercial farming. Credit and various outside inputs in the form of chemical fertilizers, pesticides, and hybrid seeds were all that was needed. There was little questioning of this strategy in the public domain.

For example, in newly independent Zimbabwe, the Department of Research and Specialist Services developed packages for the Department of Agricultural Extension to pass on to small-scale farmers. It became a nationwide effort that seemingly had good short-term results for some farmers and for the agricultural industry in general. Zimbabwe became known as the breadbasket of the region and small-scale farmers played a significant part in this.

Today there are those who still dream of industrial agriculture sweeping across Africa, and with a new weapon in its arsenal: the genetic modification of crops and animals. Certain benefactors are pouring huge amounts of money into what some call a new green revolution for Africa. Furthermore, in the name of investment, there is now a massive grab for Africa's land, the cheapest in the world. Some of this land is used for growing biofuels for the USA and Europe, and some is for growing food for the Middle and Far East.

Twenty-five years ago, what resistance there was to the spread of industrial agriculture emerged from millions of farmers continuing to use their traditional and culturally linked practices. These practices had developed over a long time and were suited to the many different biophysical and climatic environments that one finds across the eastern and southern African regions. However, in the face of the might of industrial agriculture, this became a passive and crumbling resistance. Much of the conventional agricultural world considered these practices backward. One of the most successful strategies of the advocates of industrial agriculture has been in relation to people's attitudes, positing neoliberal arguments on promises of profit and trade. Across the region, industrial agriculture is seen as "development," and traditional practices as backward and a way of surviving until the industrial methods arrive. This attitude has successfully embedded itself in many people's minds.

The difference between now and 25 years ago is that there are growing alternatives to corporate-driven, industrial agriculture. What's more, these alternatives are joining forces in more and more meaningful ways, as symbolized by the Nyéléni meeting and the launch of the Alliance for Food Sovereignty in Africa (AFSA).

The stakes are as high as they have ever been, as two fundamentally opposed approaches to food collide, one with vast resources behind it as it hangs on to its global free-market tenets, seemingly oblivious to considerations of sustainability and equity. The alternative approach gathers momentum person by person, quietly and steadily growing in what *could be* an exponential rather than a simply incremental way.

Small initiatives begin to take root in the region

Around 25 years ago, small initiatives began emerging throughout southern Africa. They offered an alternative to high-input industrial agriculture, focusing on alternative, ecologically sound farming practices. They took into account the complexity and diversity of natural systems. They were about combining production with sustainability. They were about healthy, nutritious food, and they were about conserving water, without which there would be little food. They were about the soil. They were about the environment. These initiatives looked at food in all its fullness.

These approaches were also about the complexity and diversity of social systems that the conventional, technology-transfer approaches had largely ignored. As part of understanding and working with social complexities, they recognized the critical role of indigenous knowledge and practices. Their starting point was to build on what people were already doing. In the context of their time, this was a revelation.

An illustration from Zimbabwe

Chikukwa ward, in Chimanimani District, is the smallest communal area in Zimbabwe, with six villages and a population of around 7,000. Located in the foothills of the Chimanimani Mountains, many of the small farms in Chikukwa are on steeply sloping land. During the 1980s, in line with the well-meaning but misguided government policies for smallholder farmers, many farmers in Chikukwa shifted to maize monocropping, with coffee as a cash crop. Despite the promotion of contour ridges (which actually have a downhill gradient of 0.5%) and planting of gum trees, soil erosion in the Chikukwa hills grew worse.

In 1990, concerned residents of Chikukwa asked Fambidzanai Permaculture Center, a recently established Zimbabwean NGO promoting sustainable agriculture, to run a training workshop in Chikukwa on managing land productively and sustainably. Following the workshop, the residents formed Friends of the

Land (Shamwari dze Nyika) and began to implement practices to safeguard their land while using it productively.

During a follow-up workshop the next year, there was a huge rainstorm that led to mini-landslides and terrible soil erosion. This jolted the community into action, and they formed the Chikukwa Ecological Land Use Community Trust (CELUCT) to raise funds and facilitate a number of learning activities. Gradually over the next decade, the Chikukwa community reversed the downward environmental spiral and recovered their land, largely by reducing run-off and, thus, soil erosion. Nearly all rainwater entered the ground. Springs returned and production diversified. CELUCT also established a community training center to accommodate people from outside the area, with farmers' fields as the main training sites.

Reaching out into the whole district

During the late 1990s, farmers, traditional leaders, and others throughout Chimanimani District requested CELUCT to come and work with them, having learned what was happening in Chikukwa. This was not possible, and, instead, a new organization called TSURO-of-Chimanimani came into existence. *Tsuro*, which in Shona, the local language, means rabbit, is the acronym for Towards Sustainable Use of Resources Organization.

TSURO began operating in 2000 throughout the Chimanimani district. Many people traveled to Chikukwa to participate in workshops, for which Chikukwa farmers provided training input. TSURO is now a well-established part of both Chimanimani and Zimbabwean civil society. It comprises a trust, a membership association made up of TSURO village groups from 188 villages, and a private company, owned by TSURO, which spearheads the marketing of organic produce. Like any organization, it has its challenges—one of which is to continue raising the funds needed to undertake its many activities.

Right now, in 2010, CELUCT and TSURO are working with the Zimbabwe Organic Producers and Promoters Association

(ZOPPA); African Farmers Organic Research and Training (AfFOResT), another Zimbabwean NGO; and Find Your Feet, a British NGO, in a joint program (Find Your Feet 2011). This program complements the strengths of each organization and will focus on learning and innovation among farmers, as well as improving marketing linkages.

Linking NGO initiatives

As in the example above, many of the initiatives offering an alternative to industrial, externally driven agriculture have become national or community-based NGOs. They each have their own way of doing things, sometimes specializing in a particular cropping method, like biointensive agriculture. Others have specialized in land-use design (e.g., permaculture). Others refer more generally to low external-input agriculture. Many simply use the term *sustainable agriculture* and draw on knowledge and ideas from a variety of sources.

Fairly early on, in the early 1990s, some of these local NGOs started to link up. They recognized from an early stage that they needed to think beyond their own setup and particular approach. They understood that if they were to be a force to be reckoned with, they had to learn from each other. They also noted that they were employing people who had received conventional agricultural and extension training. Such people needed to be re-equipped with alternatives in terms of agriculture as well as with new ways of working with and relating to communities.

Formation of the PELUM association

In 1995, this linking up led to the formation of the PELUM Association by representatives from 25 NGOs in East and southern Africa (PELUM 2011). PELUM stands for Participatory Ecological Land Use Management, which, though a bit of a mouthful, clearly states what's involved: people, ecology, and using the land. The initial focus of the association was building each other's capacities through sharing and learning processes. Gradually, the membership grew and the country chapters, or networks, took root. Today there are over 200 member organizations in 10 countries across the region, with active networks in each.

Bringing in advocacy

While the initial focus was teaching sustainable agricultural practices to the farmers involved, some of the NGOs began to see that their work was more than this. PELUM-Tanzania led the way in the region, becoming involved with influencing policy. This was no doubt due to their close relationship with the small-scale farmers' movement for sustainable agriculture (called MVIWATA) in Tanzania. PELUM-Kenya, with the threat of GMOs looming, led the way in lobbying against GMOs, in partnership with the Africa Biodiversity Network. At a regional level, PELUM Association policy expanded its focus to include advocacy as well as capacity building. PELUM, as an association of local NGOs working with small-scale farmers, began to participate in various advocacy-related fora, both regionally and internationally.

Birth of a regional farmers' voice

In 2002, the PELUM Association enabled 300 farmers from across the region to attend the World Summit on Sustainable Development, held in Johannesburg, South Africa. These farmers initiated the Eastern and Southern African Farmers' Forum (ESAFF) as a farmers' voice in the region (ESAFF 2011). PELUM and ESAFF work closely together, with ESAFF challenging PELUM on a number of issues. ESAFF has enabled PELUM and its members to define PELUM's role more clearly, recognizing that the most effective advocacy comes from farmers themselves and not NGOs on behalf of farmers.

The tension between an educational service and advocacy

In 2003, I was part of an evaluation team for the PELUM Association. What clearly came into view during the exercise was the growing divide among NGO members over the development of sustainable agricultural practices with farmers, and the need and potential for the association to be more involved in influencing policy change, and thus address the agrarian conditions for sustainability. Many of the organizations still saw the association mostly as a source of capacity building for themselves.

This tension remains very real today. I see this in the strategic thinking and planning work that I do with a number of community-based NGOs, most of whom are members of PELUM. They are only now beginning to see themselves as part of a much bigger movement, which is about change at all levels of society.

Other evolving networks

More recently, and growing out of the work to promote sustainable agriculture, a number of organic-farming associations have come into existence. Their focus is on developing the organic-food market, and local standards and appropriate ways to ensure certification of these standards. These associations have become particularly active in East Africa, where they have established East African organic standards. Southern Africa is now following suit.

Where Are We Now?

Now, 25 years later, there are numerous initiatives that emphasize building on local practices. Much experience has been gained both from what has worked and what has not. On the technical front, what started out 25 years ago in various small trials and projects has become a set of techniques and practices that cover most situations—that enable farmers to produce substantially while simultaneously improving their environment. This also enables them to think for themselves and make decisions that are applicable to the environment as it changes around them.

There are sustainable practices to produce a lot of food from small, intensively managed areas, and sustainable practices to produce well from more extensive, rain-fed cropping areas, the latter coming under the term *conservation farming*. In both these instances, more and more farmers recognize the immense value of well-made compost. Intercropping also plays a significant part.

One important development, particularly for drier, seasonal-rainfall environments that cover immense areas of the region, has been the recognition of the potential to use livestock to heal the land. Led by the work of the Africa Center for Holistic Management, the approach, in brief, uses cattle, goats and sheep to simulate what huge herds of wild animals like buffalo and wildebeest have been doing for millennia. Being so counterintuitive, this has faced a lot of resistance, but slowly and surely this is now taking root.

The food sovereignty movement in East and southern Africa: What is to be done?

Twenty-five years ago, we were thinking about getting off the ground. We thought about small projects and individual farmers, anything to get moving. Now our thinking must be more strategic and our visions much bigger. What were once small independent initiatives have become a worldwide movement around the recently articulated concept of food sovereignty.

Years ago, we held workshops to promote sustainable agriculture that lasted two days, a week, sometimes as long as two weeks. The workshop might involve sharing how to make compost, exploring water-harvesting techniques, designing land-use systems that maximize beneficial connections between elements, showing how to manage pests without resorting to chemical pesticides, or a mixture of all of these. There were also workshops on relating to communities in a more participatory way, using a range of visual methods. These workshops were always short and geared toward going out and doing something afterward.

Over time, organizations in Kenya such as Manor House, Baraka Agricultural College, and the Kenya Institute of Organic Farming initiated longer courses. There was also PELUM-Zimbabwe's "college without walls," which ran three two-year sandwich courses in the late 1990s and the first decade of the 2000s. These longer training programs were based on the realization that learning about all of the dimensions of sustainable agriculture linked to community development requires far more than a two-week workshop.

The farmer-to-farmer approach has grown strong within the

Food Sovereignty in Africa

A Conversation with Diamantino Nnampossa

Food sovereignty is increasingly under attack in Africa. One thing holds true in almost all African nations, and that is the condition of poverty, of malnutrition, of an exploitation rooted first in colonialism and secondly in neoliberalism. The African continent is a victim, has been a victim for many years of economic models that serve the interests of the Global North. With structural adjustment programs of the 1980s and '90s, large areas of the African continent stopped producing due to lack of support from the state. Through the privatization of the banks, public companies, and public services, neoliberal policies forced the state to sell off its public resources. After 20 years and the devastating failure of these structural adjustment programs, we Africans need new ideas that can help us overcome poverty.

Nongovernmental organizations (NGOs) have many good food sovereignty initiatives, but they are dispersed, and also there is very little political content to their work. This is why it is important for the farmer organizations in these countries to begin by making small, local changes. Rural Africa is complex. It has high levels of illiteracy and the political conciousness is still very low. So, the task of training and political education is extremely important with farmers. Our plan is to strengthen the producer-farmer base.

We need to take responsibility together, in all forums of the world, to guarantee a life of dignity for all citizens on the African, Asian, European and American continents. Especially the African farmers. They have the right to produce their own food, they have the right to control their own markets, and they have a right to protect their own agriculture. This is fundamental. This is the only way to bring the African continent out of poverty and suffering.

Full Article at: http://www.foodmovementsunite.com/addenda/Nnampossa.html

sustainable agricultural movement. The PELUM Association held a workshop in Tanzania in 1997, to share experiences from Central America. To date, the main emphasis of this approach has been on farmers as resource people, often to replace government extensionists. These extensionists have not only become scarce, following the privatization of extension services, but continue to follow conventional thinking despite its unsustainable agro-inputs.

As we look ahead 25 years into the future, we must think about education in a much more ambitious way. I see three key strands to this.

Community-based adult education

The first strand relates to adult education, with a focus on developing the farmer-to-farmer learning approach. This goes much further than seeing farmers as extension agents for NGO training programs. It is about embedding learning within communities via a range of activities, including recognizing farmer-innovators and helping to share their innovations.

This strand includes a variety of farmer-related adult education. As well as spreading practical farming information, it is about enabling farmers to influence government policies and access decision-making fora. It is also about farmers staying up to date with trends so that they are aware of what is happening and can make their own decisions. One of the crimes of industrial agriculture is that it consistently makes decisions *for* farmers all the time. Farmers play little, if any, part in the deliberations. Only when farmers have consistent access to information related to farming and food will they be able to participate meaningfully in how it unfolds. This is a key element of food sovereignty.

Food sovereignty in schools

The second educational strand relates to primary and secondary schools. This is about bringing all issues relating to food sovereignty into schools and thus into the minds of the next generations. We must recognize that schools can be a food resource for communities; they can be an area of experimentation in gardening, a model for productive and diverse land use. There is also great potential in integrating food sovereignty issues throughout the curriculum. A budding regional program called RSCOPE, which grew out of in-country initiatives, has this ambitious outlook in mind (RSCOPE 2010). Whereas 25 years ago, the aim was to help some schools design and use their land better and be an example to other schools and the communities around them, the aim of RSCOPE is a transformation of the education system by mainstreaming issues of food sovereignty.

Colleges of the future

The third strand concerns tertiary education. Here the strategy is to develop university-level courses that cover the various dimensions of food sovereignty. Universities wield, and are likely to continue to wield, enormous influence. It is essential that any food sovereignty movement in East and southern Africa participates fully in tertiary education. Already there is substantial work on curriculum development relating to food sovereignty issues. The ultimate goal is to develop these into a variety of courses.

New courses, and even new colleges or universities, can bring a fresh approach to the way in which learning occurs. Student achievements at the University of Development Studies in Tamale, Ghana, or the Faculty of Agriculture in Cuamba, northern Mozambique, are possible because these are new universities with a fresh outlook, that don't remain wedded to the past. In both these institutions, students spend substantial time living among rural communities and experiencing rural life directly, as part of their agricultural learning.

The rapid spread of the Internet brings with it many opportunities for the development of distance education programs. Whereas 25 years ago, those initiating the development of sustainable agriculture in East and southern Africa could not see beyond short-term training courses in order to get things going, the new advent of connectivity across regions and continents invites us to think about appropriate, full-length educational programs.

The role of NGOs

Local NGOs have played a pioneering role in the promotion of sustainable agriculture in the region, and thus have forged the groundwork for a viable food sovereignty movement. As highlighted above, their role has focused primarily on an adult education service, with some moves toward influencing policy.

Throughout the next 25 years, as the food sovereignty movement gains momentum, NGOs will need to examine their roles thoroughly in a new light. The issue is no longer one of projects in this or that community, but of playing a significant part in a social movement. It's about being part of joined-up efforts. There may be valuable projects, and there will of course be work with communities, but the question is one of attitude within the context of a changing game being played out. More than ever, NGOs will need to be on their toes and be willing to change.

Everything NGOs do in the future must relate to shifting toward a food sovereignty movement. First, of course, they must come to clearly understand throughout their ranks what food sovereignty means and how it is evolving. In planning their work with communities, they must consider how they can contribute to the development of this movement. What information can they provide, for example? What networks can they play an active part in? Most important, how can they enhance the voice of farmers, so that they can truly be heard?

All this will mean a deeper and more constant understanding of what is happening in specific contexts at various levels. It will mean further developing the art of strategic thinking. It also means that NGO staff must see their work as much more than a job, as is usually already the case in the more effective NGOs. When we see ourselves as part of a collective and growing movement, we can begin to influence the rules of the game.

Different approaches to food are part of a power struggle, and NGOs must understand the nature of this struggle and clearly define their roles within it. None of this is easy for a civil society that is still young. It is therefore important that older, more established NGOs play a leadership role. They have a responsibility to guide and mentor the younger organizations. They also have a responsibility to forge new directions. The NGO model as it is now, dependent on international funding and with a tendency to become urban and office-bound, needs to move on, developing better ways to serve and empower farmers.

One area that I believe is crucial relates to how NGOs make decisions. This starts with their outlook. We must shift away from a problem/poverty focus that tends to treat symptoms. Instead, we must make decisions geared toward a desired and sustainable outcome, based on an appreciation of strengths and resources in hand. This process of decision making, in recognition of social and ecological complexity, will examine possibilities from many angles.

Another crucial area is that of gathering evidence and documentation. Stories of success will add momentum to the movement. Failures or difficulties will enable learning. NGOs have a history of weakness in this area and should work closely with willing scientists.

Farmers' organizations and farmers as advocates

Farmers' organizations are going to play a pivotal part in the development of a food sovereignty movement in East and southern Africa. This has become very clear over the last 25 years. At the same time, there has been a poor record of NGO work with farmers' organizations. The African continent is littered with the skeletons of infant farmers' organizations, created by NGOs, that failed to root themselves in the interests of the farmers themselves. While it has generally been common practice to design projects that include the establishment of farmers' organizations, these organizations have often been about delivering an NGO project and not about the establishment of a long-lasting farmers' organization that grows out of the farmers' own volition. Multinational NGOs have been particularly guilty of this.

Fortunately, there has recently been a growing realization of the dangers of this approach, and a shift toward working with existing structures and setups established by the farmers and communities themselves. It is essential that this trend continue and spread. At the same time, there must be a deepening of organizational development (OD) work with farmers' organizations. There has been much of this OD work in the private and civil sectors, but it has been seriously lacking at the community level, especially in relation to farmers' organizations. The training is generally too superficial. Over the next 25 years, we must forge OD from a vision of vibrant farmers' organizations.

As more and more genuinely rooted and vibrant farmers' organizations establish themselves, so will the voice of the farmers grow. This voice will enable more effective value adding and marketing, spurring constant deliberation of issues. This deliberation, at the heart of food sovereignty, will in turn influence policies at all levels, from local to national to global, and across sectors.

Farmer-based research

Many of the strategic directions I am outlining overlap. This is particularly the case with research and education. Conventional, corporate-driven agriculture continues to carry out research at stations and develop "packages" for farmers, as it has done for decades. There is little farmer involvement. Recently, because of criticism related to this, there has been some token involvement; testing certain seed varieties before release is one example of this.

A food sovereignty approach to research will require that farmers be involved at every stage. In fact, professional researchers are there to serve farmers, a reality that often eludes researcher and farmer alike. Fortunately, some have demonstrated an understanding of this. Many NGOs in East Africa have been working with farmers who undertake their own trials; the Best Bet approach, which I have come across often in western Kenya, is one example of this. In this approach, a farmers' group uses a small piece of land belonging to one of their members to test new practices like intercropping. They watch and monitor these trials closely and make up their own minds as to what works best for them. In Zimbabwe, AfFOResT worked closely in the late 1990s with farmers in the Zambezi Valley to try out different intercropping systems for organic cotton. They combined these trials with using the farmer field-school approach, which is an excellent tried-and-tested way of involving farmers in their own research. Many farmer field schools have successfully operated in the region over the last 10–15 years.

The next 25 years should see a widespread movement toward communities running their own farming trials, supported by scientists and field workers. This could be combined with the establishment of a strategically sited community "Ecolabs," who have the kind of equipment that enables farmers to learn more about the science of ecofarming. For example, the role of microorganisms is being recognized more and more. It is important that farmers have a chance to see these microorganisms for themselves and understand their enormous benefit. Only then will they become relevant.

Public awareness and consumer organizations

In East and southern Africa, there is a dearth of prominent consumer organizations related to food sovereignty. The overriding focus has been on sustainable production, with a few tentative moves toward involving consumers. It is very important that the next 25 years see the establishment of vibrant consumer organizations across the region. The only one I know of presently is in Ethiopia. I'm sure there are more, but they are not nearly as conspicuous as they should be. These consumer organizations will make the wider public, particularly the fast-growing urban public, aware of food sovereignty issues and how they relate to their own health and well-being. Healthy food produced in healthy ecosystems that adequately compensate growers for their work is a solidarity issue among all ordinary people, rural and urban.

Linked to this is urban farming. This practice is overcoming the colonial prejudices that tried to keep food production out of urban spaces. Combined with good land-use design and intensive production practices, urban farming is already taking off and will form an increasingly important part of the urban landscape over the next 25 years, supporting the food and livelihood needs of growing migrant populations.

Financing the development of the food sovereignty movement

As indicated earlier, local NGOs have played a significant part in the promotion of sustainable agriculture over the last 25 years in East and southern Africa. Nearly all the financing of their work has come from countries in the North, mostly via European and American NGOs and church organizations. Given today's financial crisis and uncertainty, this is a tenuous position; breaking free of these confines is one of the biggest challenges facing local civil society, especially since the food crisis is stimulating research investment from biotech organizations that threaten to drown all other voices and subsectors. There are no easy routes to financing development work like this. It will require that local civil society organizations think strategically about this issue and continue to explore different avenues.

Nevertheless, despite concerns about a decrease in funds for international development, or "aid," as it is often called (a term I find misleading—is it not a failed concept?), signs indicate this source of funding will continue in the short to medium term. A sizeable minority of the public in the Global North is aware that a more equal, just, and healthy world is desirable for everyone. It is therefore essential that we continue to raise awareness. This is where food sovereignty organizations in the North can play a solidarity role with those in the South.

Civil society involved in food sovereignty in East and Southern Africa must improve its relationships with donors while seeking nontraditional sources of collaboration. There are 25 years of experience and learning that we can build on. We will learn the lessons and build strategic alliances in which there is deep-rooted understanding between partners. At the same time, there is a need to explore and move toward new relationships between Northern and Southern populations. After all, that's what funding is about: one population with surplus resources financing adult education or advocacy work in another part of the world. At present, this relationship operates through a certain infrastructure. It is my belief that we must shift and change this infrastructure and that players in the North and South should deliberately explore how this can happen.

One possible way this could occur is by establishing linkages with multinational NGOs. The constant presence of these NGOs in the region does not contribute to the development of local civil society. In some cases, it actually undermines it. A strategic move would be for local NGOs to develop partnerships with these multinational NGOs, so that the latter continue to raise funds for the former to implement work within their own civil societies. This would be a step forward and represent a growth in solidarity between North and South.

However, it is critical that civil society organizations also find ways to raise their own funds. There has not been a good record of this to date, but a great deal has been learned, and there are some small successes to build on. How can the food sovereignty movement in East and southern Africa, in conjunction with the global movement, start to use its economies of scale to attract funding, create markets and revenue streams, and establish more independent funding? This is the critical question. There are excellent examples of this in other parts of the world; for example, in Bangladesh, the civil society organization BRAC raises around 75% of its financing from social enterprises (BRAC 2011).

AFSA and continuing to join up our work

All of the above depends on a continual process of knitting together civil society efforts around food sovereignty. We need to build a strong food sovereignty movement both locally and regionally that is also part of the worldwide movement. This will create the social and political force that is necessary to overcome the domination of the present corporate food regime, based as it is on industrial agriculture and processing and the false solutions of the Green Revolution. The industrial agriculture complex has a history of destroying the environment, enriching multinational corporations, and contributing to an increasingly unjust and unfair world.

Without the power of a strong social movement, we will never create the social, political, and economic conditions for sustainable agriculture. For this we need powerful alliances like AFSA. That is why its launch, growing out of La Via Campesina's Nyéléni meeting in 2007, is momentous. Now is the time and the opportunity to bring practitioners, advocates, and consumers together in a powerful way. This is a political stance because it is about people standing and acting together. AFSA is a potential bridge between local learning and global food politics. It is a catalyst, a rallying call, the next chapter in an ongoing story—a chapter that is much more strategic than before.

On the one hand, AFSA will be a light-on-its-feet think tank, challenging us all with its bird's-eye view. What is emerging? What will be the next strategic focus? What's working and can be shared? What conceptual and practical shifts are needed? How could, or should, the story unfold?

On the other hand, AFSA must draw on activities at various levels to develop policy drives. At the heart of policies are values. AFSA will bring together and articulate the food sovereignty values that resonate across Africa. Food has ecological, social, economic, cultural, spiritual, and technological dimensions to it. Industrial agriculture has reduced it to a commodity and the term "food security," rendering it devoid of meaning. Food is much more, and that's why all those within the umbrella of AFSA and the food sovereignty movement in general will rescue and revive it.

Based on clearly articulated values, AFSA will play a significant part in influencing policy and helping others to influence policy as well. At the very least, we need policies that put in place farmer-driven research toward ecologically sound production, policies that represent small farmers' voices across the continent, policies that ensure land-tenure continuity for small farmers, and policies that institute education at all levels in sustainable African farming and marketing—not an education based on conditions and practices from other parts of the world. We need policies that also enable urban areas to produce high quantities of healthy food, policies that link health issues to food production and processing, and policies that direct financing to the kind of farming that benefits primarily the land and lives of small farmers and gardeners and their communities—not corporations.

Perhaps the greatest challenge for AFSA will be to hold the area that separates industrial agriculture from all that food sovereignty promotes. AFSA must take an inclusive approach in order to build the strong movement we need, while sticking to its values. The danger, which so often surfaces when resistance is involved, is the possibility for infighting around the details and losing momentum. Aziz Choudry, a longtime activist and researcher, formerly the organizer of GATT Watchdog and currently an assistant professor at the McGill University in Montreal, says:

> Activism is bound to always face lots of contradictions and ambiguities, but this should not be a barrier to building more linkages. There is a clear need to build alliances that respect people's different situations and worldviews. The most significant and effective struggles are happening in movements that are grounded in local contexts but connected to global perspectives. This is difficult, non-glamorous movement building work that, incrementally, is creating spaces where power can be challenged. We rarely hear about these struggles, but they are where hope for the future lies (GRAIN 2010).

Works Cited

BRAC. About BRAC Social Enterprises. Accessed January 29, 2011. http://www.brac.net/content/about-brac-social-enterprises.

Declaration of Nyéléni. 2007. World Forum on Food Sovereignty. Accessed January 29, 2011. http://www.nyeleni.org/?lang=en.

ESAFF. Accessed January 29, 2011. http://www.esaff.org/.

Find Your Feet. Accessed May 1, 2011. http://www.find-your-feet.org.

Food First. "Alliance for Food Sovereignty in Africa (AFSA) Challenges African Leaders on Climate Change." Food First/ Institute for Food and Development Policy. November 28, 2009. http://www.foodfirst.org/en/node/2670.

GRAIN (Genetic Resources Action International). Seedling. 10/07/2010. A ccessed June 6, 2010. http://www.grain.org/seedling_files/seed-10-07.pdf

PELUM. Accessed January 29, 2011. http://www.pelumrd.org.

RSCOPE. Accessed January 29, 2011/ http://www.seedingschools.org.

Chapter Six

TRANSFORMING NGO ROLES TO MAKE FOOD SOVEREIGNTY A REALITY

By Fatou Batta, Steve Brescia, Peter Gubbels, Bern Guri, Cantave Jean-Baptiste, and Steve Sherwood of Groundswell International

WE KNOW THAT agroecological farming works for family farmers in Africa, Latin America, and Asia, and that family farmers represent the great majority of the world's people who face extreme poverty and lack adequate food. We know farmers need net-beneficial relationships to markets, and that it is necessary to create policies that support rather than undercut the well-being of rural communities.

How can nongovernmental organizations (NGOs) best contribute to making food sovereignty a reality? We will attempt to answer such questions through drawing from our practical experience with prominent food movements in Haiti, Ecuador, Burkina Faso, and Ghana.

Food sovereignty is unquestionably a powerful framework for organizing responses to the dysfunctional global agrifoods system. Linking local, democratic control and decision making to the foundational economic activity of all societies—producing and eating food—is a powerful agent of change on many levels. But what does the concept mean to a farmer watching for rain while planting seeds on an inaccessible mountainside in Haiti; a peasant organization in Burkina Faso looking for strategies to shorten the hungry season; or potato farmers in Ecuador trying to escape their dependence on expensive fertilizers and toxic pesticides? Bern Guri of Ghana says food sovereignty in his country means "people having access to sufficient food and nutrition, but also being able to have control over their own food system, producing what they eat and eating what they produce." If NGOs are to play a useful role in these people's lives, they must develop practical strategies to help them achieve their personal goals.

Peter Gubbels provides some broader analysis:

> For many years, Ghana has been seen as a model country, because it has been greatly influenced by the policies of the World Bank and other proponents of the neoliberal economic paradigm. As a result, Ghana largely neglected its own food security. There is an alarming trend toward large-scale export crops such as exotic fresh vegetables, pineapple, agrofuel, and mangoes, and corporate control of resources for production. It is well documented that Ghana's policies provide insufficient protection against imports from countries with generous subsidy regimes, resulting in Ghana importing a significant proportion of its staple rice and basic grains. This left the Ghanaian population—particularly the poor, most of whom are rural people—highly exposed to the spiral in world prices during 2008. The food crisis did finally stimulate the Ghanaian government to abandon its noninterventionist position and start investing in agriculture. Unfortunately, Ghana's response is to modernize agriculture and increase productivity based mostly on a "green revolution" approach, which has been tried many times in Ghana and never succeeded.

In this context, Gubbels believes that "working for food sovereignty in Ghana means promoting agroecological methods of production, enhancing biodiversity and local control of seeds, ensuring fair prices for small-scale farmers, strengthening markets and processing links between peasant producers of healthy local food and urban consumers. It also means organizing and advocating for an alternative to green-revolution approaches based on the principle of 'African solutions to African problems.'"

So what are some practical strategies that NGOs can use to achieve these goals?

- Transform the role of NGOs in the intended participants' lives
- Promote farmer innovation and agroecological production
- Expand territory for agroecology
- Build productive alliances with farmers' movements and strengthen their bases
- Advocate policy reform without neglecting crucial practices
- Take advantage of new opportunities (health, urban-rural linkages, and climate change).

None of these strategies speak to the quick-fix mentality of many donor agencies, multinational corporations, and politicians. To thrive, they must be rooted in local contexts and led by local people.

Rethinking and Transforming the Role of NGOs

Many NGOs are doing valuable work, but having worked with NGOs for decades, we are aware of their limitations and problems. The case of Haiti is often illustrative of the problems with development assistance and the roles played by NGOs.

On January 12, 2010, a catastrophic earthquake in Haiti devastated Port-au-Prince and surrounding cities. One major reason the devastation was so great in these cities is that Haiti's underlying rural foundation had been greatly weakened. Rural farming communities had been systematically drained of resources for centuries by bad economics and politics, both domestic and international. The most prominent resource that remains in Haiti is the Haitian people themselves (the majority of whom is still rural and farming based), their tenacity and capacity for organized action. Years of rural migration to cities with inadequate infrastructure, housing, or jobs contributed to over 250,000 people being killed in the earthquake. In the days following the earthquake, 600,000 displaced people fled back to the countryside—at least temporarily. Peasant communities and organizations responded by receiving, housing, and feeding these people, depleting their own limited food and seed stocks. Partnership for Local Development (Partenariat pour le Développement Local; PDL) helped channel a small portion of the emergency assistance following the earthquake to these rural organizations. They used the resources efficiently for both short-term relief and investment in the long-term solution of revitalizing devastated rural areas, as a foundation for Haiti's future. "We are strengthening local peasant organizations so that they can be actors in leading their own development," says Cantave Jean-Baptiste. "Over the last 20 to 30 years, we have seen that strong peasant organizations adopting agroecological farming, improving local seeds, soil management, and so on, are key in Haiti, and have been making long-term improvements in rural communities."

Despite the proficiency of peasant organizations, the clear need for decentralization in Haiti and the demonstrated effectiveness of agroecological approaches, peasant organizations have largely been left out of shaping or implementing plans for Haiti's recovery. Drawn up by international experts and a Haitian government with limited capacity or credibility with its own people, recovery plans only pay lip service to priorities like promoting agriculture and domestic food production, supporting family farming, involving peasant organizations, and decentralizing the country. In practice, the implementation of these plans defaults to typical top-down interventions heavily biased toward the importation of what for poor farmers are expensive technologies. Commenting on the plans, Jean-Baptiste says, "I see seeds, fertilizers, and tractors, but I don't see farmers. Where are the farmers?" The plans are generally implemented by putting contracts out to bid to development companies and NGOs.

This example from Haiti illustrates the most frequent role NGOs play: implementing contracts for plans rural people have neither designed nor agreed to. The world of official development assistance usually either misses the opportunity to work with rural people and farmers' organizations, or works in opposition to their interests. Most aid money is strongly influenced by the paradigm of industrial agriculture, which seeks to extend its model and inputs to small-scale farming. NGOs too frequently end up being the implementers of this agenda, and fit into service-delivery and relief categories. Few strengthen the capacity of local people and organizations to transform their economies sustainably, and few support agroecological farming.

So what should NGOs do and not do? In Haiti, "NGOs can play a technical role in supporting agroecological production, but they should also strengthen the capacity of local organizations to carry out their own development," says Jean-Baptiste. "NGOs commonly respond to their headquarters and not to communities, and therefore have limited interest in coordinating with each other to learn what works."

"In Ecuador, NGOs are a mixed bag," says Steve Sherwood. "NGOs have become donor driven and project driven. This has limited their ability to be responsive to local needs and be creative. Project-based giving has hurt NGO effectiveness. For EkoRural, we try to keep our role as small as possible. We don't try to find solutions for local communities, but find what is working, ask good questions, support local creative ideas, and facilitate exchange to help these to grow. Having limited financial resources forces us to be responsible, and rely on local people's leadership."

"In Ghana and in most of Africa, most NGOs have a technical focus rather than linking to social movements. NGOs are supposed to be waging the war on food insecurity, but most are isolated entities," says Bern Guri. "There are reasons: they are struggling to survive and responding to donor demands, rather than working in coalition. In CIKOD we suffer from these challenges as well. We have our vision. I may think that one important way to encourage food sovereignty is through improved practices in communities, but some funders only support advocacy. This can create frustration."

NGOs that want to strengthen community-driven change to create a healthy agricultural and food system and economy need to find ways to meet these challenges through:

- being responsive to the interests of rural communities and organizations, rather than to funders, while developing greater downward accountability to those communities;
- critically analyzing what kind of farming works for small-scale farmers in the developing world;
- connecting effective community-level action to wider policy reforms;
- developing alternative sources of funding when the donor community demonstrates limited willingness to invest in agroecological and farmer-led approaches with a track record of success; and
- seeing their role primarily as strengthening local capacity for sustained change and working themselves out of a job, rather than delivering services.

Promoting Farmer Innovation and Agroecological Production

For NGOs sincerely intent on transforming rural communities, the starting point must be the people—not a technology, a particular crop, or even a specific sector per se (agriculture, health, microfinance, etc.). The question must be: How can we support rural people in generating well-being and overcoming poverty? We've learned much from decades of collective experience and trial and error in thousands of villages in Africa, Latin America, and Asia. The key lessons are that authentic, community-led development is always holistic and based on strong local capacity, and that agroecological farming is a vital means for rural people to improve their lives. An increasing number of evaluations and studies are affirming similar conclusions (McIntyre et al. 2009).

Why is agroecological farming important for small-scale farmers? The primary reason is that it works. Farmers own the process—managing, adapting, and creating it. It improves their lives—often reversing declines while doubling or even tripling production. The majority of small-scale farming work is now globally done by women, and as Fatou Batta says, "Women are often leaders in adopting agroecological practices because it is accessible, meets their needs, and can lessen their workload. And in addition to farming, women are also the real link connecting improved production to better family consumption and nutrition." Agroecological farming is economically, environmentally, and culturally sustainable. It strengthens communities, local leadership (including women), and local organizations. It improves the natural-resource base that people depend on. Agroecological farming is an economic strategy for the poorest people to overcome hunger—to produce and eat a diverse and adequate amount of food, and to generate income.

By contrast, over the last 50 years we've seen countless programs focused on high-external input agriculture do the opposite. We are reminded of some farmers we visited in Guatemala's highlands a couple of years ago. They had become contract farmers producing broccoli for "a company," renting land each season and buying seeds, fertilizer, and pesticides as prescribed. As we stood in their plot, our feet planted on soil devoid of organic matter, looking at broccoli plants dwarfed by a disease they did not understand, one farmer said, "At first it was a miracle, but now we are enslaved by this system. We make less money every year, and we have to calculate each year if we should plant again, or migrate. We are trapped. I would tell other farmers to farm another way."

NGOs working to combat this trap created by many aid programs, implement a strategy that supports small-scale farmers, local organizations, and wider movements to learn about, innovate, and expand the use of agroecological farming as a practical alternative to improve their lives. "We can't transform the global food system unless farmers are able to expand the practice of sustainable farming and increase their control over how they farm," says Peter Gubbels.

Agroecological farming means more than continuing the old ways or simply training men and women through a new package of sustainable practices and technologies. Some farmers practice both traditional techniques that are sustainable (seed saving, crop diversity, etc.) and those that are no longer sustainable (slash and burn). Others adopt elements of industrial agriculture and reliance on external inputs. Farmers do what they think works for them, and we've seen both types benefit from transitions to more agroecological farming methods that are appropriate to their conditions: small plots, marginal and barely farmable land, fragile ecosystems, degraded soils, and isolation from services and markets.

What have we found are the most effective strategies for promoting farmer innovation and agroecological farming? In our experience, successful strategies revolve around allowing farmers to discover what works for them and spreading these alternatives through their social networks. Key methodologies NGOs can employ include

- farmer experimentation and innovation—on their own farms;
- farmers identifying key limiting factors and testing a small number of alternatives to see what works;
- strengthening farmer-to-farmer networks to spread successful practices;
- focusing on seeds, soils, and water—managing, improving, and making the best use of these local resources; and
- cultivating diverse, integrated farms.

While specific technologies will evolve with local conditions and opportunity costs, as our colleague Roland Bunch has written, farmers' capacity to innovate must remain a constant theme (Bunch and Lopez 1994). This means people engaging in the creative, evolving act of farming, and avoiding dependence on external inputs that uproot that capacity.

"In Burkina Faso, industrialized agriculture is expanding and being promoted by some political leaders," notes Fatou Batta. "Village-level farmers are not aware that when they sell land or give production rights for jatropha for biofuels, they and their children and grandchildren lose access. We've seen the importance of supporting people to learn what works locally, supporting agroecological approaches and resisting the pressure of some donors to promote a high external-input approach, a quick fix, instead of listening to local people. Our evaluations have shown that the *zai* technique for water and soil conservation, nitrogen fixing trees, and short-cycle seeds result in 50% to 120% increases in production. It is very high risk for farmers to depend on external inputs and distant markets, and drives them into poverty and off the land."

In Ecuador, as in many countries, the majority of those managing family farms are women. Organizations like EkoRural are helping them strengthen local seed systems through farmer field schools. Through discovery-based learning processes, they are supporting farmers adapting to the effects of climate change, such as depleted groundwater and altered rainfall patterns. Farmers measure the value of rainwater lost from their roofs and fields, and "harvest" it in simple storage tanks, for future use, and, most importantly, in their fields—as increasing organic matter in the soil allows for more water to be stored in it. The result is a positive cycle of increased productivity and innovation, and significant improvements in family well-being, nutrition, and income. Steve Sherwood says that in Carchi, a potato-producing region heavily dependent on dangerous and highly toxic pesticides, "farmers have learned to maintain and increase their production using agroecological practices while reducing or eliminating the use of expensive and dangerous pesticides."

Expanding Territory for Agroecology

Bern Guri says that in Ghana, "isolated examples of small-scale farmer agroecological production exist, but the government of Ghana doesn't see what small farmers are doing as relevant, because they are focused on larger farmers. They see small-scale farmers as holding back production. We need to shine a light on the successful examples, but also create a market. To do this, we identify capacities that farmers already have for agroecological farming and strengthen them and spread them. We work to document and disseminate the good work already happening so people know that alternatives exist."

A common critique of agroecological practices is that they always appear to work for a small number of farmers but are never widely adopted. Why is this often the case? There are a few possible reasons: (a) farmers are not aware of agroecological alternatives; (b) they are aware of them, but aren't convinced that they work, or believe that something else works better; (c) incentives (economic, environmental, social, or psychological) push them toward certain ways of farming. NGOs must work together and with farmers to overcome these constraints and develop more effective strategies to scale agroecological practices across communities and regions—to expand the territory for agroecology and healthy local food economies.

The question of whether or not agroecological farming works for small-scale farmers in the developing world is perhaps the easiest to address. As noted above, broad experience along with a growing body of research and evidence demonstrates that it works for them on multiple levels. Even proponents of industrialized agriculture usually accept agroecology's success on a small scale, but argue it is not viable on a larger one. Yet many farmers in the developing world are already adopting and practicing agroecological farming, and the only incentive they have to do so is that it brings them benefits—more food, less cost, an improved environment, healthier families and communities, greater resilience to shocks, and so on. While there are a powerful set of actors with a major economic self-interest in promoting the sale of their agricultural inputs and technologies, the same is not true of agroecology. The only incentives for external actors to promote agroecology are social—reducing poverty and creating a more inhabitable planet.

So how can we spread awareness of agroecological farming among rural communities? What strategies can make these practices more effective, and how can we create incentives for using them, so that the territory for agroecological farming and local food economies is expanded?

- **Farmer-to-farmer and community-to-community** Nothing convinces farmers like showing them how they themselves can increase production on their own farms. Visiting farmers who have succeeded in the same conditions is a powerful motivator for them to learn as well. We have long employed these farmer-to-farmer strategies to reach a critical mass (30%–40%) of innovative farmers in a community. Once such a critical mass is reached, successful practices tend to spread to others over time. The same strategy can be applied to communities, as Cantave Jean-Baptiste notes: "We can also facilitate communities to visit and learn from each other, and to develop plans for action together."
- **Capacity strengthening** Managing, sustaining, and further scaling these farming methods inescapably requires strong local organizations and networks of rural people. For NGOs, this implies some combination of working with existing community-based organizations and strengthening their capacities for self-management. While NGOs often get stuck in a cycle of delivering services, some have developed strategies for strengthening the capacity of community-based organizations. In Haiti, PDL has created a highly effective approach to strengthening local peasant organizations (which typically group together 15–30 villages). The foundation of these intervillage associations are *gwoupman*, solidarity groups of 8–15 people who work, learn, and apply agroecological

practices together, pool savings for loan funds, and create and manage both seeds banks and tool banks. Representatives are elected from the *gwoupman* to form committees that coordinate activities within and across communities. This allows them to take on challenges individual farmers can't manage on their own (e.g., controlling free-grazing animals) and increase their ability to access markets and advocate for health services and schools. In this way, PDL is strengthening the social infrastructure needed to scale sustainable farming and build local economies (they are currently working with nine local peasant organizations representing over 148,000 people). Cantave Jean-Baptiste adds, "We are now working to create networks of these local peasant organizations so that they can work together and support each other. As an NGO, we need to facilitate communities and local organizations to learn from each other, work together and lead their own process of development."

- **Action-learning networks** In eastern Burkina Faso, isolated examples of successful agroecological approaches exist, even under the very difficult Sahelian conditions that are currently being exacerbated by the pressure of growing population. But the spread of these "islands of success" through a wider adoption of agroecological practices is constrained by a lack of sharing and coordination among the local NGOs and community-based organizations responsible for the work. And there is no significant effort by the government or major donors to promote, invest in, or spread these alternatives. In response, a new network of local organizations is emerging in the region to facilitate the sharing of knowledge of successful strategies and define action plans to replicate them.

Peter Gubbels believes that "we should invest strongly in farmer-to-farmer and community-to-community learning and exchange, particularly within agroecological zones where the climatic conditions, crops, and farming systems are similar. Without practical examples of how successful agroecological methods can be taken to a much wider scale, it will be difficult to make a compelling case to other NGOs, the Ghanaian public, and policy makers that this is a viable alternative to the industrial, export-oriented, green revolution approach to agriculture."

Building Alliances with Farmers' Movements

Many have criticized NGOs for focusing on technical approaches to supporting agricultural (even agroecological) development while failing to fully collaborate with farmers' movements in promoting food sovereignty and changing policy. It is often a fair critique.

Farmers are important social actors in rural people's organizations, articulating the interests of their members and giving them political voice. NGOs must identify effective means of supporting and strengthening them as autonomous political entities. Unfortunately, NGOs can easily lose sight of this and put themselves in the center of policy debates. As Bern Guri notes, "NGOs should try to strengthen farmers' voices in the political process and not replace them."

It must be emphasized that NGOs and farmers' organizations are diverse, and neither type of organization is immune to the challenges that tend to face any organization. Developing and implementing strategies that are effective, broadening and renewing leadership, remaining driven by values and mission, or avoiding overly centralized decision making and power structures are just a few of these. Both must focus on promoting the interests of rural people and achieving food sovereignty, and there is ample opportunity for them to collaborate.

We have been involved with a number of NGO efforts over the years to collaborate with farmers' organizations and movements, particularly in Latin American and Caribbean countries including Guatemala, Honduras, Nicaragua, Haiti, Ecuador, Bolivia, and Peru. The goal has generally been to strengthen local agroecological pilot initiatives that can be scaled throughout existing networks. Unfortunately, these efforts often fall short of their potential impact. Farmers' movements have in some cases demonstrated commitment to their genuine need for land rights and political influence, but expressed little interest in sustainable farming methodologies. Meanwhile, NGOs, despite their best efforts, have failed to build adequate trust in negotiating their role in strengthening the community-level base of broader movements. Clearly, both political voice and appropriate farming methods are vital. Changes in political policy are necessary; one can't farm without access to land or if production is undercut by subsidized imports. But even with land and policy supports, successful farming still requires a locally-led process of innovation for productivity and sustainability.

There is a need for NGOs and farmers' movements to engage in honest dialogue: to examine common interests and what each brings to the table, to look for win-win opportunities that can be gained through collaboration, and to develop trust. This often happens by starting with small, concrete initiatives.

"Most leaders of Ecuador's indigenous movements have worked closely with NGOs over the years," says Steve Sherwood. "There has been a lot of productive collaboration. But many NGOs have become project driven. And even many indigenous leaders have become urbanized. As they have gained power; they need to live in cities and get involved in politics. This has weakened the indigenous movements in some ways. Both indigenous leaders and NGOs need to get replugged into rural families and communities."

"In Burkina Faso, limited movements exist to promote agroecology," notes Fatou Batta. "Groups tend to be working in isolation. An agroecology platform does exist in Burkina, but it is not very strong. Those social movements tend to be stronger in Mali. So in Burkina we need to support efforts to pull things together and show the viability of these alternatives."

In the words of Cantave Jean-Baptiste, "In Haiti, we are strengthening the base. We need to strengthen local peasants and their organizations to assume the roles of actors in leading their own development. We also facilitate them in strengthening networks across many communities, and to connect to the wider peasant-movement organizations."

Most of the local peasant organizations in Haiti belong to wider peasant movements and networks. While these networks play a vital role in Haiti's development, they would be further strengthened by greater participation from their base groups, and better two-way flow of communication between the base and peasant-network leaders. Jean-Baptiste notes that "sometimes the peasant organizations also need to do a better job of communicating with their own base. For example, while peasant movements were protesting and burning hybrid Monsanto seeds in Haiti in June 2010, I visited some of their base groups that had received some of those same seeds from the AID-supported program. The farmers did not have adequate information about what to do with the seeds, or what the impact would be if they became dependent on hybrids. Some of the farmers were even eating the pesticide-covered seeds as grain, which is dangerous."

Peter Gubbels observes that "most members of farmers' organizations in Ghana are larger-scale commercial farmers. They are organized in associations around the production and marketing of specific commodities like rice, tomatoes, poultry, and cotton, and advocate for policies affecting their particular commodity. This includes seeking government subsidies for inputs, agricultural research, and for trade regulations that prevent dumping or subsidized imports. Yet these groups are not representative of the mass of semisubsistence peasant farmers, men and women, who are mostly illiterate, and who practice traditional agriculture with hand tools. Most members of the influential farmer organizations are oriented to agribusiness or industrial methods of production. So while their advocacy for trade regulations that prevent dumping, and for government subsidies for inputs and agricultural research is compatible with food sovereignty, their approach to production and sustainability often is not."

Bern Guri believes there is an opportunity in Ghana to strengthen a movement from the bottom up. "We need to work through indigenous institutions, such as chieftaincies, which are closest to the people, and are legitimate and respected." Chieftaincies have strong influence with rural people, and control community land; therefore, they have the potential to change communities' attitudes, to promote agroecological innovations and revalorize local seeds and food crops. "We can support these indigenous institutions to build a mass movement. NGOs need to have the capacities to do that."

Advocating Policy Reform without Neglecting Crucial Practices

Social movements in Ecuador are powerful and have had notable success in reforming major policies. Indigenous people represent the majority of Ecuador's population and are effectively organized into local, second-level, and national organizations. They have demonstrated their political power by shutting down the country through strikes, and even bringing down governments. The indigenous movements form a significant arm of the social movement for progressive reforms to the constitution and other laws.

"There have been important policy achievements in Ecuador," says Steve Sherwood of EkoRural, "such as the passage of a food sovereignty law and a law to eliminate the use of highly toxic pesticides. The Colectivo de Agroecologia, which EkoRural is a part of, is a network that brings these actors together, including linking urban consumers with small-scale producers. They have helped to draft and shape the food sovereignty law. It was an important landmark for us to see that it was possible to influence policy, but it also showed us the limitations of policy. Policy is just on paper. Practice depends on what people do."

Companies representing the interests of industrialized agriculture still manage to insert themselves into the process and highjack the debate. Recent history has proven that changing policies alone is not enough. We are supportive and are trying to influence policy.

But if we do not influence what people and families actually do, how they produce and consume, then we will not have achieved enough.

Peter Gubbels highlights the challenges created by the Ghanaian government's allowance of subsidized food to be dumped in the country. "This has to change if there is to be a people-centered food system in Ghana! Strengthening local food systems first requires both fair and protective trade policies that enable local farmers to sell their food production to Ghanaian consumers. There are many low-cost, economically feasible policies that Ghana could promote to improve the production, marketing, and processing of local food crops. For example, government policies could support decentralized milling of locally grown rice to meet consumer expectations. They could make appropriate credit and small-scale irrigation accessible to semisubsistence peasant producers for dry-season gardening. Appropriate forms of crop insurance for small-scale farmers could be developed. Ghana should also explore systems to ensure that peasant farmers obtain a reasonable price for food crops, and promote marketing at the local and national levels."

Taking Advantage of New Opportunities: Health, Urban-Rural Linkages, and Climate Change

"In rural Burkina," says Fatou Batta, "we promote community-managed grain banks to increase food security. Farmers sell at a better price and have local access to less expensive food during the hungry season." As in many countries, farmers typically sell to middlemen after harvest, when the price is lowest, and then need to buy back from those same middlemen at the time the price is highest. Community grain banks help them break the cycle. In Haiti, farmers typically pay exorbitant annual interest rates of 250%–500% on loans from local moneylenders—just to obtain seeds and tools to plant at the beginning of the agricultural cycle. We support them in setting up their own savings and credit groups, seed banks, and tool banks to liberate themselves from this debt trap.

But we need to go beyond helping rural communities stop the drain of resources, and support them in achieving prosperity. As Steve Sherwood and his colleagues in Ecuador have discovered, "We need to think about agriculture and food as an integrated system. The choices we make about how we eat are key. Working only on agriculture has excluded farmers from the wealth of urban people. Ecuadorians spend six to eight billion dollars a year on food. How can we bring this consumer wealth to bear on transforming rural landscapes?" Urban consumers, many of whom are low income and need better access to healthy food at reasonable costs, can be the "funders" of small-scale agroecological farming production.

To promote this, EkoRural and other organizations in Ecuador have been supporting the emerging *canastas comunitarias* movement: a type of community-supported agriculture arrangement. Low-income, urban consumers have formed groups to buy food wholesale and thereby lower its costs, and are now directly connecting to small-scale farmers and building buying relationships with them. "We found an example that works and expanded on it," says Sherwood. "This started with one group. We worked with them to think critically about nutrition per dollar spent, and gradually about how to promote the rural landscapes and communities that we want through what we buy and eat. We promoted critical thinking through cross visits and building relationships between urban and rural people. This has now grown into a *canastas* movement that has gone from a few groups to all major cities in Ecuador."

In Ghana, Bern Guri notes that "we need to demonstrate the health implications of our traditional foods. When the Director of Health in Ghana bought local millet porridge on the street and emphasized the health benefits in the media, the market for these products boomed. The government could promote this. They could create a policy that 1% of all food served in restaurants must come from traditional food. Right now restaurant food is imported. We could target urban consumers, youth and school feeding programs, linking them to traditionally grown foods from small scale-farmers. It would help promote young people's tastes for these local foods."

The need to respond to climate change presents another opportunity for pushing back against industrialized agriculture. "We can link efforts to adapt to climate change to the promotion of people-centered food systems," says Peter Gubbels. "This is possible because most solutions to adapt to climate change in rural communities require agroecological approaches, rather than those based on industrial agriculture." Emerging payment mechanisms for carbon sequestration for soil high in organic matter and agroforestry may provide opportunities and additional incentives for farmers.

Conclusion

Haiti's earthquake occurred in a just a few minutes but caused a scale of destruction and death that shocked the world. The global food earthquake has been playing out over a longer time frame, and it impacts each context in different ways. Tremors like food-price increases periodically expose the scale of the devastation around the world to those who may not be living it daily. The Haitian and global tragedies have similar roots—centuries of marginalization and exploitation of rural people via economic and political systems that don't serve their interests. This has weakened the very building blocks upon which any strong society must be built—producing healthy food and communities, regenerating the land and environment, and allowing people to participate democratically in shaping their future.

In Haiti, as well as in Ecuador, Burkina Faso, Ghana, and around the world, people are working to rebuild a healthy foundation from the bottom up. There is a great need and opportunity for people to come together to continue to build on these efforts and to meet the challenges of the moment. Along with family farmers, rural, indigenous, and urban people's organizations, governments and donors, technicians and political activists, people in the Global South and North, NGOs also have an important contribution to make. Yet NGOs must continue to challenge themselves to focus on people-led development and to promote practical strategies that work: support for local innovation and sustainable, appropriate farming; strengthening the capacities of local leaders and organizations to manage their own change processes; strengthening local food economies; spreading successful alternatives via farmer-to-farmer and community-to-community sharing; and creating alliances with wider social movements to influence policy. We all need to find ways to contribute to reconnecting healthy farming, healthy eating, and healthy democracy. This is the shared task of building food sovereignty together.

Works Cited

Bunch, Roland, and Gabino Lopez. 1994. "Soil Recuperation in Central America: Measuring the Impact Four to Forty Years After Intervention." Accessed March 13, 2011. http://rolandbunch.com/articles/.

McIntyre, Beverly D., Hans R. Herren, Judi Wakhungu, and Robert T. Watson, eds. 2009. "Sustaining African Agriculture: Organic Agriculture." In *Agriculture at a Crossroads: The International Assessment of Agricultural Knowledge, Science and Technology for Development.* Washington, DC: Island Press.

PART TWO:

CONSUMERS, LABOR AND FOOD JUSTICE

Chapter Seven

SURVIVAL PENDING REVOLUTION: WHAT THE BLACK PANTHERS CAN TEACH THE U.S. FOOD MOVEMENT[1]

By Raj Patel

OVER THE PAST DECADE, the US food movement has grown to become a potent force for social change and, precisely because of its success, the movement now is being called to shore up the status quo. Revisiting some radical roots suggests ways that the food movement can end hunger in America, rather than becoming just another impermanent band-aid for poverty.

Critical thinking about and organizing around food in the United States aren't new— Frances Moore Lappé's work gave rise to the institute publishing this book, Food First, nearly 40 years ago (Lappe 1971; Lappé and Collins 1977). Food scares and diet fads shaped US public consciousness about food through the 1980s and '90s. But I suspect it's no accident that the movement grew after the terrorist attacks of September 11, 2001. As Michael Pollan (2010) notes in his piece "Food Movement Rising" in the *New York Review of Books*:

> [It] makes sense that food and farming should become a locus of attention for Americans disenchanted with consumer capitalism. Food is the place in daily life where corporatization can be most vividly felt: think about the homogenization of taste and experience represented by fast food. By the same token, food offers us one of the shortest, most appealing paths out of the corporate labyrinth, and into the sheer diversity of local flavors, varieties, and characters on offer at the farmers' market.

---

[1]. I've been schooled by many people in the writing of this paper, most of all by Kiilu Nyasha, Michael William Doyle, Gayatri Menon, and Eric Holt-Giménez.

To be sure, a food movement predated 9/11—the National Family Farm Coalition was founded in 1986; environmentalists had been taking on Monsanto, spurred by Rachel Carson's 1962 *Silent Spring*; and the history of American disenchantment with capitalism is as old as the nation itself (Zinn 2003). The American Revolution wouldn't have happened but for the actions of merchants protesting the terms of trade for tea. (Schlesinger 1917). Yet it was only with the criminalization of dissent, with the increased difficulty of confronting corporate capitalism through other politics, and with the fear coursing through the veins of the US public after 2001 that the movement's strands were more tightly woven together. Under the Bush regime, environmentalists, social justice campaigners, anticapitalists, and organic foodies found a government, media, and general public far less responsive than a decade before. Membership of umbrella groups like the Community Food Security Coalition has swelled, with a proliferation of food organizations, consultants, academics, and activist groups throughout the US. Under these circumstances, a new generation of activists was drawn into the movement. What is particularly striking—and although I haven't anything but anecdotal evidence to offer in support of this view, I'd be happy to bet—is the relative youth of those moving into the movement. He may blush, but Josh Viertel—president of Slow Food USA and contributor to this volume—is in his early 30s, and that's not an accident. He's a prodigious leader in a new generation of activists like Oakland's Brahm Ahmadi and Nikki Henderson, who have organized and advanced food justice in the US during the first decade of the 21st century.

Part of the success of the movement has been its largely nondenominational, big-tent approach, committed to the idea that food should be a pleasure available to all, and that, above all, food is a domain in which something can and ought swiftly to be *done*.[2] Indeed, it's the very success in community farms, gardens, feeding programs, kitchens, and food banks that has

[2] In part, this vision has roots that can be traced to the outsized anarcho-Marxist organizing that produced Slow Food (Andrews 2008). See also Pew Research Center (2010).

helped recruit more and more people to a movement that seems to offer the transcendence of "old politics" so earnestly cashed in by the Obama campaign in its first election run.

Yet it's the movement's practical success that puts it in a precarious position today. At the time of this writing, hunger is its highest levels in a generation (Nord et al. 2010)—50.2 million Americans are food insecure, and one-third of female-headed households are food insecure. At the same time, food prices are rising, unemployment remains stubbornly high, and a Republican Congress has ambitions to amputate social programs from the body of government in the name of fighting inflation (Patel 2011). In the resulting vacuum, community organizations have been pressed, much to government's approval, into the business of service provision. As Suzi Leather remarked of a similar period in the UK government's history:

> It is easy to see the appeal of the community development approach for the present administration: it smacks of the self-help ethos, involves vanishingly small resources and can be encouraged without at the same time having to admit to the existence of poverty. (Leather 1996, 47–48)

To inoculate ourselves against the dangers of being co-opted into the very food system we have spent a decade criticizing, we need politics. Two instant caveats, though. First, merely talking about the politics of the modern food system isn't sufficient to prevent the movement's energy from being dissipated while dealing with the "dignified emergency"[3] of increasing hunger. History is littered with all-night activist conversations about the root causes of hunger, with little change to believe in the morning after. Second, a call to talk about capitalism in the food system isn't a call for a single totalitarian politics to which all must subscribe. Every US social movement, from abolition to the Tea Party, has drawn on an assortment of sometimes contradictory political positions.

[3]. This is a phrase I have learned, and embrace, from Nick Saul's work at The Stop. For an example of how this thinking informs community organizing in dignified emergency here, see Scharf, Levkoe, and Saul (2010).

The problem is that the food movement's ideological pantry is rarely raided, and despite a rich history, there's not nearly enough talk about it. By food politics, I don't just mean the kinds of interaction between state and private sector presented by Marion Nestle in her fine dissection of the food industrial complex (Nestle 2002). I'm referring to politics as an ideology, as a positive system of beliefs, analytical principles, and values that informs practice (Badiou 2005; Hall 1996; Rancière 2007). And of these systems of politics, there seems insufficient praxis. Perhaps the origins of the food movement, in politically embattled times, is to blame for a certain ideological quietism. But whatever the food movement's genealogy, its future needn't be hostage to the past.

Activist Anim Steele's (2010) work, drawing on the civil rights movement, dips into movement history a little, but it's worth remembering that the civil rights movement itself was hardly homogenous. Its demands for political and civic rights were nested in further demands for economic and social rights—a recognition that Martin Luther King (1967) himself made explicit toward the end of his life:

> One day we must ask the question, "Why are there forty million poor people in America?" And when you begin to ask that question, you are raising questions about the economic system, about a broader distribution of wealth. When you ask that question, you begin to question the capitalistic economy. And I'm simply saying that more and more, we've got to begin to ask questions about the whole society. We are called upon to help the discouraged beggars in life's market place. But one day we must come to see that an edifice which produces beggars needs restructuring. It means that questions must be raised. You see, my friends, when you deal with this, you begin to ask the question, "Who owns the oil?" You begin to ask the question, "Who owns the iron ore?" You begin to ask the question, "Why is it that people have to pay water bills in a world that is two-thirds water?" These are questions that must be asked.

King moved through a wide field of politics, in which talk about the failures of capitalism was part of popular discourse, and which King himself began to embrace more fully toward the end of this life. Certainly, the civil rights movement addressed issues of hunger. The day after King's assassination, the NAACP subverted a USDA press conference at the USDA, announcing their intention to sue the government for its failure to bring school lunches into compliance with civil rights legislation (Levine 2008, 136). But it's another movement from which I'd like to learn, one that both thrived and was destroyed because it addressed issues around food, and that offers something surprising and powerful for our political imaginations today.

The Black Panthers Feed the World

Although poverty had been worse in the late 1960s, and although poverty would worsen again, African Americans were as disproportionately represented among the hungry as they had ever been. One of the constants of post–World War II US life has been that African American income has consistently stayed at around 60% of white household income (DeNavas-Walt, Proctor, and Smith 2009). The federal government's persistent refusal to address poverty in African American communities was compounded in the 1960s by an ongoing criminalization of poor, urban African Americans by local and state police, with attendant and systematic police violence against black men. It was the encounter with this "police logic" that spurred two students at Merritt College in Oakland, Huey Newton and Bobby Seale, to launch the Black Panther Party for Self Defense—later shortened to the Blank Panther Party (BPP) (Cleaver and Katsiaficas 2001; Hilliard and Cole 1993; Rancière 1998; Seale 1970; Singh 1998). The party initially organized and armed themselves to monitor the Oakland police in December 1966, opening their office in Oakland in January 1967 (Seale 1970).

The party soon expanded its ambit beyond police surveillance, dropping "for Self Defense" from its name and, through dialogue with community members, setting up a range of community service programs. By 1968, the most successful of these, the Breakfast for Children Program, was up and running in the Bay Area and Seattle (Abron 1998; Newton, Hilliard, and Weise 2002, 15).

Youth and Food Justice:
Lessons from the Civil Rights Movement

By Anim Steel

Improving the health of our youth will require a transformation of our food system. This in turn will require strong social movements capable of creating the political will to truly transform how we grow, buy, prepare, and eat food. Lessons from the civil rights era of the 1960s suggest a way that today's food justice movement can organize. In particular, a new, youth-led, multiracial coalition could unleash the voice and energy of those with the most to gain from transforming the food system—young people.

The political disenfranchisement addressed by the civil rights movement in the 1960s, and the cheap, unhealthy food plaguing our underserved communities both reflect structural inequities that marginalize people of color. We can't change the food system by simply changing the tastes and attitudes of regular people any more than the civil rights movement could end segregation without the 1964 Civil Rights Act. Beyond the personal, these transformations require political, economic, and cultural changes. Just as with the civil rights movement, transformation needs to be local, national, and international. Social movements will play a deciding role in creating the political will for change just as they did with civil rights.

To become a strong national force, the food justice movement needs a youth-led organization that unifies and amplifies these disparate efforts—a modern-day food justice version of the Student Nonviolent Coordinating Campaign (SNCC). Such an organization should celebrate and encourage the diversity of local work; the best local solutions come from local communities. But it should do what local organizations often have a harder time doing: focus the national spotlight, spread innovation, involve masses of people, and harness our collective political and economic power. Such an organization should prioritize the voices of those most hurt by the system, even as it welcomes the contributions of all who care.

Youth Food Movements Unite!

Full Article at: http://www.foodmovementsunite.com/addenda/steele.html

The origins of the program aren't clear. In some Panther writings, it appears as an endogenously chosen, natural outcome of a commitment to "serve the people" (Seale 1970). Bobby Seale also suggested that he arrived at the idea through conversations with local teachers, and needed to persuade Eldridge Cleaver, the Panther's minister for information, who thought that free breakfasts were a "sissy program," but was eventually won over (Rhodes 2007, 251).

This isn't the breakfast program's only creation story, though. In his memoir, David Hilliard, the Panther's chief of staff (Hilliard and Cole 1993), recalls a donation of food given by Emmett Grogan, an activist with the Diggers in San Francisco (Grogan 2008, 475). The Diggers, a breakaway group from the San Francisco Mime Troupe performing arts group, traced their name and, in part, their politics to the 17th-century movement resisting enclosure in England (Gurney 1994). The original Diggers were stout defenders of communitarian property and collective self-government of agricultural land. The modern Diggers blended situationist[4] performance with their predecessors' agrarian communism through "events" like giving away free food. Grogan describes how the free-food giveaways in San Francisco's Panhandle district involved stepping through a bright-orange window frame called the Free Frame of Reference, so that when the hungry emerged on the other side, their frame of reference had been changed (Grogan 2008, 250). Note, incidentally, that after an initial attempt at cooking the food himself, Grogan passed on responsibility for doing it to "a half-dozen young women, a few of whom were dropouts from Antioch College, shared a large pad together on Clayton Street and volunteered to take over the cooking indefinitely" (248).

No matter who did the cooking, it's clear that these events were a well-publicized part of the 1960s Bay Area counterculture, and

4. Situationism offers a critique of the mass media under capitalism. The French intellectual Guy Debord wrote situationism's classic book, *The Society of the Spectacle* (2002), in which he argued that "[a]ll that was once directly lived has become mere representation."

it's more likely that the Panthers knew of them than not (Doyle 2011). Grogan writes of a meeting following the 1968 killing of Black Panther activist Bobby Hutton, in which Grogan, together with "Black Panther Party Chairman Bobby Seale, and Chief of Staff David Hilliard . . . began discussing a plan they had to start a Free Breakfast for Children program that would put some nourishment into the normally empty bellies of black kids before they went to school" (Grogan 2008, 474–75). It'd be tempting to chalk this tale to Grogan's literary bravura—his autobiography often plays fast and loose with the truth—but David Hilliard's discussion of Grogan corroborates some of the facts, and is worth quoting at length:

> Emmett Grogan sticks his head in the office. Emmett is the founder of the Diggers, a tribe—that's what some radicals call their groups—who organize the "street people" of the Haight into revolutionary activity. A few weeks ago, Emmett left off some bags of food his group distributes to the runaways, draft resisters and freaks who have flocked to Berkeley, turning the town into the nation's counterculture capital. We told him to put the stuff outside the office: in a few minutes people were flocking by, stocking up on onions and potatoes. Now Emmett donates the food regularly. Like the newspaper, the food serves a double purpose, providing sustenance but also functioning as an organizing tool: people enter the office when they come by, take some leaflets, sit in on an elementary PE [political education] class, talk to cadre, and exchange ideas, all part of the revolutionary ferment I have imagined when listening to Huey describe Fidel and Che in Cuba. (Hilliard and Cole 1993, 158)

What's nice about this story is, first, the piercing of racial boundaries traditionally associated with the Panthers. It turns out that—surprise!—there were interactions between different radical groups within the Bay Area, and that they learned from one another. What's also nice about it is the ability to trace the genealogy of the free breakfast programs back to radical movements to defend the commons. Finally, of course, what matters here is not just that the food was distributed—even the federal government was doing that, very poorly, as part of the school breakfast program in the 1966 Child Nutrition Act. What distinguished the Black Panther Party's food distribution was its part in a far wider vision for social change.

Part of the mechanisms of the Black Panther Party's self-defense were programs for survival, ranging from the provision of free shoes and education to land banking and the school breakfast program (Huey P. Newton Foundation and Hilliard 2008). In the provision of these services, Newton understood the ambiguities and contradictions within the programs:

All these programs satisfy the deep needs of the community but they are not solutions to our problems. That is why we call them survival programs, meaning survival pending revolution. We say that the survival program of the Black Panther Party is like the survival kit of a sailor stranded on a raft. It helps him to sustain himself until he can get completely out of that situation. So the survival programs are not answers or solutions, but they will help us to organize the community around a true analysis and understanding of their situation. When consciousness and understanding is raised to a high level then the community will seize the time and deliver themselves from the boot of their oppressors. (Huey P. Newton Foundation and Hilliard 2008, 4)

The breakfast program itself served a shifting menu, with varying degrees of success, numbers served, and outreach in the 45 different branches nationwide.[5] New York's chapters fed numbers in the hundreds, California's in the thousands. Nonetheless, the universal aspiration was for a balanced diet of fresh fruit twice a week, and always a starch of toast or grits, protein of sausage, bacon, or eggs, and a beverage of milk, juice, or hot chocolate (Huey P. Newton Foundation and Hilliard 2008, 31). In practice, the breakfasts were constrained by funds and in-kind donations.

[5.] Heynen (2009) cites his interview with Bobby Seale, in which the figure of 45 branches and 4,000 members was presented.

The FBI was keen to prove that these donations were extorted from local businesses, but despite considerable effort, failed to do so (Newton, Hilliard, and Weise 2002, 340). Meanwhile, there is now a wide consensus that, for many children, the meals were the only source of nutrition in a child's day.

Beyond the success in feeding, there was a political component to the program. The *New York Times* (Caldwell 1969) represented the breakfasts as austere "diets of food and politics" at which children recited the dour mantras of the movement: "I am a revolutionary; I love Huey P. Newton; I love Eldridge Cleaver; I love Bobby Seale; I love being a revolutionary; I feel good; off the pigs; power to the people."

In some cases, the police and FBI were successful in casting the breakfasts as not only doctrinaire, but as dangerous, with rumors circulating that the Panthers were serving poisoned food, and would rape girls if they could (Abron 1998). In one case, the Chicago police allegedly broke into a Panther feeding facility and urinated on the children's food the night before it was to be served (Heynen 2009, 414). In some places, particularly New York, those rumors took hold, and parents kept their children away from the programs. Yet a recording made at the New York breakfast program suggests that the brainwashing wasn't always successful—when a 12-year-old boy starts calling for "Fewey Hewton" to be freed, everyone felt safe enough to laugh along (KPFA and Kamen 1970, 15:30).

A more subtle understanding of the program's politics, one repeated by activists in print and interviews, is that the breakfasts were explicitly geared toward demonstrating what socialism might look like (Hilliard 2007; KPFA and Kamen 1970). In a touching moment in one testimony, a woman recalls a child's transformation, after being found filling his pockets with food and hearing that he wasn't stealing but that the food was his and would he like a bag. As Joan Kelley, national coordinator of the Black Panther Breakfast Program said, "We try to teach children not so much through indoctrination but through our practice and example about sharing and socialism" (KPFA and Kamen 1970, 6:14). By bursting the idea of food as a charity bestowed by rich to poor, setting in its place the notion that food is a right—and the suggestion that an order might be composed without private property—the act of feeding children was transformed from pacifying to revolutionary, without a single "Free Huey" passing anyone's lips.

The breakfast program was part of a suite of survival programs with explicit goals of transforming relations around private property—the vision of a land bank, for instance, called for the creations of trusts that would suspend the profit motive from land tenure, making other arrangements possible (Davis 2010). Land reform was, in turn, part of a broader political strategy, enshrined in the Panthers' Ten Point Plan, which featured "power to determine the destiny of our black and oppressed communities," "full employment," "an end to . . . robbery by the capitalists," "decent housing," "decent education," "completely free health care," and an end to war, militarism, police brutality, and, in the final point, "land, bread, housing, education, clothing, justice, peace and people's community control of modern technology." It's hard to argue that this longer vision, the goal of emancipation postponed, didn't infuse the feeding programs with a political momentum missing from common philanthropy's food banks. It was the political vision, the possibility of a different tomorrow after surviving today, that transformed the Panthers' feeding into radical social work (Bailey and Brake 1976).

Effect and Aftermath

Jesse Jackson called the breakfast program "creative and revolutionary" (Levine 2008, 139), and insofar as it survived the direct assaults on participation, and also survived the indirect weakening of the movement, it thrived. People across the country copied it. In Milwaukee, Wisconsin, Pastor Joseph Ellwanger of the Cross Lutheran Church formed the Citizens for Central City School Breakfast Program (which later became the Hunger Task Force of Milwaukee) after refusing to let the breakfast program use his church (White 1988, 90).[6] The Young Lords, a Puerto Rican youth organization, set up similar feeding programs in Chicago and New York (Judson 2003). In Austin, Texas, two African Americans started up a feeding program without the Panther's politics, but sending reports back to Panther headquarters from time to time (KPFA and Kamen 1970). At a national level, the breakfast program stoked grassroots pressure that eventually led to increase in funding for kids' food (Levine 2008, 140).

And this is not because the program had a "radical flank effect" (Haines 1984), in which organizations making infinite demands of capitalism create space for more accommodating organizations to achieve their aims (Critchley 2007). The breakfast program actually fed children. In a Senate hearing George McGovern asked the school lunch program administrator, Rodney Leonard, "whether the Panthers fed more poor children than did the state of California . . . Leonard admitted that it was 'probably true.' [Senate Select Committee, Part 11, July 9–11, 1969, 3478]" (Levine 2008, 139).

The Panthers' success in providing food also intensified the efforts to crush them. The FBI was, through its COINTELPRO program, trying to destroy the Panthers. The government found it much harder to summon popular support for its work when the Panthers were engaged in radical social work. As Ward Churchill observed, "[FBI director J. Edgar] Hoover was quite aware that it would be impossible to cast the party as merely 'a group of thugs' so long as it was meeting the daily nutritional requirements of an estimated 50,000 grade-schoolers in forty-five inner cities across the country. Rather than arguing that the government itself should deliver such a program, however, he targeted the Panthers' efforts for destruction" (Churchill 2001, 87).

Squeezed by the actions of the state, but also torn by serious internal political divisions, the Panthers buckled. Some insisted

---

6. Though he earned himself the Panthers' derision in their newspaper, the *Black Panther* (July 5, 1969, 5), where he was referred to as "a punk, racist, fascist pig preacher."

on the vision of survival pending revolution, maintaining a fidelity to the principles that founded the party. Others, particularly senior leaders, found that the state might provide an avenue for political work. So, as Nikhil Pal Singh notes:

> By the early 1970s, Panther Party leaders Bobby Seale and Elaine Brown sought effective black public input into metropolitan resource distribution in a credible, grassroots political campaign for Oakland City Council. Once again, even in its most self-consciously revolutionary idiom, black power unfolded both within and against the American grain. (Singh 2004, 210). The breakfast program, like the party itself,[7] began to fade, with activists burning out, selling out, dropping out, and being assassinated. Indeed, the breakfast program was part of the split in the party. After a fake letter was sent by the FBI in 1971 to Eldridge Cleaver, who was then in exile in Algiers, Cleaver attacked the Panthers' central committee, arguing that the breakfast program was reformist (Newton, Hilliard, and Weise 2002, 358).

But the program has an important legacy. Not only was it responsible for creating what today might be called a "temporary autonomous zone" (Bey 2003), for instigating real "school meal revolutions" (as opposed to the kind shown on today's TV), and for embarrassing the federal government into taking child nutrition seriously, but—at least in some cases—it involved a transformation within the domain in which the Panthers have consistently been considered remiss: gender.

In an important and thoughtful paper, geographer Nik Heynen (2009) presents a series of interviews with women who were part

---

[7]. The most obtuse explanation for the movement's demise comes from the world of political science. David O'Brien (1975), using Mancur Olsen's (1971) logic of collective action, explains the BPP breakfast program's decline, thus: "The key mistake in the breakfast program was that BPP failed to realize that by opening participation to anyone who wanted to come, they would encourage free riders, who duly turned up, causing them to lose revenue and, ultimately, fail." Even more important are the explanations that look into the political and cultural shifts of the 1970s and the party's own political mistakes (see Booker 1998; Johnson 1998).

of the BPP's feeding programs in the 1970s. One activist cited by Heynen[8] spoke, like many others, of the lengthy discussions and dialogues around gender, and the lengths to which the Panthers earnestly but inconsequentially paid lip service to questions of gender equality[9] and then said:

> You could have a thousand dialogues on gender issues and you would have never gotten that result faster than you did by saying look, if you love these children, if you love your people, you better get your ass up and start working in that breakfast program. (413)

It was the active participation in the program that transformed gender relations, not merely the talking about it.

This vision of gender transformation isn't, however, widely shared. When I asked one activist whose work was based in New Haven about Heynen's ideas, she was unimpressed. She wasn't alone—many of the women who were part of the Panthers engaged not because of the enlightened gender praxis, but despite it (LeBlanc-Ernest 1998; Matthews 1998; Nyasha 1990). Indeed, the only way in which many women were taken seriously within the movement was not because of equality over the cooking range, but because they were armed. For some women within the Black Panther Party, power grew out of the barrel of a gun.

But it's not inconceivable that, among the dozens of Panther chapters, even if women have reported the persistence of patriarchy, this sexist bubble might also have been punctured by moving men into kitchens and onto serving lines for children.

---

8. Heynen unfortunately omits details of the specific program with which the woman was associated.

9. Abu-Jamal (2001), for instance, observes the lengthy debates. But the debates are entirely compatible with the persistence of sexism in the day-to-day operation of the organization, as witnessed by women within the party.

Conclusion

The Black Panthers' vision for radical change is one from which the food movement today might benefit. The Panthers understood that while the needs of the hungry were real, and deserved immediate attention, those needs could only ultimately be banished by a far more radical transformation than the government was ready to provide. Political education was, the Panthers knew, vital to understanding the reasons behind their hunger. So they read Mao, Frantz Fanon, and Marx. They also knew that the combination of political education and effective action made them dangerous, turning them into enemies of a status quo that produced hunger. Hence the massive government-sponsored attempts to murder their example, and parade its body as a warning to those whose hearts might harbor similar hopes.

Yet the Panthers' example remains important for today's food movement. Clearly, it's difficult to balance the desire to recruit a broad movement under a single banner, and the need to broach the potentially divisive subject of capitalism. You can find this tension within the notion of "food sovereignty" that guides the international peasant movement, La Via Campesina. Their definition of "food sovereignty" has changed over time (Patel, 2010), though it is at heart a call for political equality at every level of the food system, so that decisions about the food system might be made democratically.

With an organizational structure as diverse as La Via Campesina's, vagueness is politically expedient. In a movement peopled with landed peasants and landless workers, any talk about "the means of production" is fractious—some folk in La Via Campesina have land and are reluctant to talk about giving it up—even if talking about all of this might provide more political focus. Food sovereignty is, from the outset, an idea built on postponing certain difficult political discussions to another day—just as long as everyone gets a say in what a new food system might look like.

Precisely because equality in political participation has to come first, the one conversation that can't be avoided or postponed is the one about gender. Although questions about unequal ownership may be punted to tomorrow, the consequences of gender inequality need to be addressed today. Hence a recently launched campaign confronting violence against women, which itself is the product of hard conversations, and concerted organizing by women within La Via Campesina (2011). The campaign stretches not only to domestic violence, but the structural violence of poverty, i.e., to those inequities magnified by capitalism.

For La Via Campesina, some of the most powerfully transformative and practical parts of a theory about global change in the food system come from actual gendered fights for the future of food. The Black Panthers' struggles for survival may not yet have brought the revolution, but at least they saw the scale of change needed so that hunger might finally be banished in our communities. And in the US today, the group most likely to be food insecure are households headed by women. It's possible to explain why this is so—why women are paid less than men, why hunger flourishes among the poor, and why capitalism will not willingly provide food to those unable to afford it.

In providing these explanations, and organizing effective actions to address inequity, we will make the food movement more threatening to the powerful. That sounds frightening, but every movement that has ever accomplished social change—whether the civil rights movement, the Indian independence movement, or indeed the global justice movement—has put the demands of justice ahead of the need to accommodate oppressive thinking. Instead, such movements have been armed with radical ideas for a better future, in which all people are possessed of dignity, and able to govern themselves. The Black Panther Party's vision of a world where all children are fed, where food, healthcare, education, access to land, and housing and clothes are rights and not privileges is a vision that can and should spark the food movement today. Inspired by their example, and learning the lessons from their experience, we can dream beyond the limitations imposed by capitalism, of a world in which hunger is, for the first time, a specter of the past.

Works Cited

Abron, JoNina M. 1998. "'Serving the People': The Survival Programs of the Black Panther Party." In Charles E. Jones, *The Black Panther Party [Reconsidered]*. 177–92.

Abu-Jamal, Mumia. 2001. "A Life in the Party: An Historical and Retrospective Examination of the Projections and Legacies of the Black Panther Party." In Cleaver Katsiaficas, *Liberation, Imagination and the Black Panther Party*. 40–50.

Andrews, G. 2008. *The Slow Food Story: Politics and Pleasure*. Montreal: McGill-Queen's University Press.

Badiou, A. 2005. *Metapolitics*. J. Barker, trans. London: Verso.

Bailey, Roy V., and M. Brake. 1976. *Radical Social Work*. London: Edward Arnold.

Bey, Hakim. 2003. T.A.Z.: *The Temporary Autonomous Zone, Ontological Anarchy, Poetic Terrorism*. 2nd ed. Brooklyn, NY: [Great Britain]: Autonomedia.

Booker, Chris 1998. "Lumpenization: A Critical Error of The Black Panther Party." In Jones, Charles E. *The Black Panther Party Reconsidered*. 337–62. Baltimore: Black Classic Press.

Caldwell, Earl. 1969. "Black Panthers Serving Youngsters a Diet of Food and Politics." *New York Times* (1923–Current). 57. http://spiderbites.nytimes.com/pay_1969/articles_1969_06_00002.html.

Churchill, Ward 2001. "To Disrupt, Discredit and Destroy: The FBI's Secret War against the Black Panther Party." In Katheryn Cleaver, and G. Katsiaficas, eds. *Liberation, Imagination and the Black Panther Party: A New Look at the Panthers and Their Legacy*. New York/London: Routledge. 78–117.

Cleaver, Katheryn and G. N. Katsiaficas, eds. 2001. *Liberation, Imagination and the Black Panther Party: A New Look at the Panthers and Their Legacy*. New York/London: Routledge.

Critchley, Simon. 2007. *Infinitely Demanding: Ethics of Commitment, Politics of Resistance*. London: Verso.

Davis, John E. 2010. "Origins and Evolution of the Community Land Trust in the United States." In *The Community Land Trust Reader.* 3–47, edited by J. E. Davis. Cambridge, MA: Lincoln Institute of Land Policy.

Debord, Guy. (2002) 2005. *The Society of the Spectacle*. Ken Knabb, trans. Berkeley: Bureau of Public Secrets. Accessed May 27, 2011. http://www.bopsecrets.org/SI/debord/.

DeNavas-Walt, Carmen, B. D. Proctor, and J. C. Smith. 2009. *Income, Poverty, and Health Insurance Coverage in the United States: 2008*. Retrieved from http://www.census.gov/prod/2009pubs/p60-236.pdf.

Doyle, Michael W. 2011. "Conviviality and Perspicacity: Evaluating Sixties Communitarianism." In *West of Eden: Communes and Utopia in Northern California*, Iain Boal, J. Stone, M. Watts and C. Winslow. eds. Oakland, CA: PM Press.

Grady-Willis, Winston A. 1998. "The Black Panther Party: State Repression and Political Prisoners." In Jones, *The Black Panther Party [Reconsidered]*. 363–90. Baltimore: Black Classic Press.

Grogan, Emmett. 2008. *Ringolevio: A Life Played for Keeps.* New York: New York Review Books. Distributed by Signature Book Services.

Gurney, John. 1994. "Gerrard Winstanley and the Digger Movement in Walton and Cobham." *The Historical Journal* 37 (4): 775–802.

Haines, Herbert H. 1984. "Black radicalization and the funding of civil rights: 1957–1970." *Social Problems.* 32 (1): 31–43.

Hall, Stuart. 1996. "The Problem of Ideology: Marxism without Guarantees." In Stuart Hall: *Critical Dialogues in Cultural Studies*. edited by D. Morley and K.-H. Chen. London: Routledge.

Heynen, Nik. 2009. "Bending the Bars of Empire from Every Ghetto for Survival: The Black Panther Party's Radical Antihunger Politics of Social Reproduction and Scale." *Annals of the Association of American Geographers.* 99 (2): 406–22.

Hilliard, David, ed. 2007. *The Black Panther: Intercommunal News Service.* New York: Simon & Schuster.

Hilliard, D., and L. Cole. 1993. *This Side of Glory: The Autobiography of David Hilliard and the Story of the Black Panther Party.* Boston: Little, Brown.

Huey P. Newton Foundation, and D. Hilliard, eds. 2008. *The Black Panther Party Service to the People Programs.* Albuquerque: University of New Mexico Press.

Johnson, Ollie A. III. 1998. "Explaining the Demise of the Black Panther Party: The Role of Internal Factors." In Jones, *The Black Panther Party [Reconsidered].* 391–414.

Jones, Charles E., ed. 1998. *The Black Panther Party [Reconsidered].* Baltimore: Black Classic Press.

Jeffreys, Judson. 2003. "From Gang-Bangers to Urban Revolutionaries: The Young Lords of Chicago." *Journal of the Illinois State Historical Society.* 96 (3): 288–304.

King, Martin L., Jr. 1967. "Where Do We Go from Here." Speech given at the Southern Christian Leadership Conference. Atlanta, Georgia. August 16. *Famous Speeches and Speech Topics.* Accessed May 27, 2011. http://www.famous-speeches-and-speech-topics.info/martin-luther-king-speeches/martin-luther-king-speech-where-do-we-go-from-here.htm.

KPFA, and J. Kamen. 1970. "Revolution for Breakfast" (radio program). *Broadcast on August 14, at KPFA Radio*, Berkeley, CA.

Lappé, Frances M. 1971. *Diet for a Small Planet.* New York: Ballantine.

Lappé, Frances M., and J. Collins. 1977. *Food First: Beyond the Myth of Scarcity.* Boston: Houghton Mifflin.

La Via Campesina. 2011. "Message from Dakar: Peasants Confront Land Grabs, Violence against Women, and AGRA." Posted February 2, 2011. http://rajpatel.org/2011/02/07/message-from-dakar-peasants-against-land-grabs-violence-against-women-and-agra/.

Leather, Suzi. 1996. *The Making of Modern Malnutrition: An Overview of Food Poverty in the UK*. London: Caroline Walker Trust.

LeBlanc-Ernest, A. D. 1998. "The Most Qualified Person to Handle the Job: Black Panther Party Women, 1966–1982." In Jones, Charles E. *The Black Panther Party [Reconsidered]*. 305–36.

Levine, Susan. 2008. *School Lunch Politics: The Surprising History of America's Favorite Welfare Program*. Princeton/ Woodstock, UK: Princeton University Press.

Matthews, Tracye A. 1998. "No One Ever Asks, What a Man's Role in the Revolution is: Gender and the Politics of the Black Panther Party, 1966–1971." In Jones, Charles E., ed. 1998. *The Black Panther Party [Reconsidered]*. Baltimore: Black Classic Press. 267–304.

Nestle, Marion. 2002. Food Politics: *How the Food Industry Influences Nutrition and Health*. Berkeley: University of California Press.

Newton, Huey P., D. Hilliard, and D. Weise. 2002. *The Huey P. Newton Reader* (A Seven Stories Press 1st ed.). New York: Seven Stories Press.

Nord, Mark, A. Coleman-Jensen, M. Andrews, and S. Carlso. 2010. "Household Food Security in the United States, 2009." *USDA Economic Research Service*. PDF. Retrieved from http://www.ers.usda.gov/Publications/ERR108/ERR108.pdf.

Nyasha, K. 1990. Public Meeting. Boalt Hall, Berkeley: University of California Berkeley.

O'Brien, David J. 1975. *Neighborhood Organization and Interest-Group Processes*. Princeton/London: Princeton University Press.

Olson, Mancur. 1971. *The Logic of Collective Action: Public Goods and the Theory of Groups*. Cambridge, MA: Harvard University Press.

Patel, Raj. 2011. "That witch, inflation, hurts us more without protection." *The Guardian*. January 19.

———. 2010. “Food Sovereignty: An Introduction.” *Journal of Peasant Studies*. (37) 3, 663-672.

Pew Research Center. 2010. *The Millennials: A Portrait of Generation Next. Confident. Connected. Open to Change.* Washington DC: Pew Research Center.

Pollan, Michael. 2010. “Food Movement Rising.” *New York Review of Books*. June 10.

Rancière, Jacques. 1998. *Disagreement: Politics and Philosophy*. Minneapolis: University of Minnesota Press.

———. 2007. On the Shores of Politics. London: Verso.

Rhodes, Jane. 2007. *Framing the Black Panthers: The Spectacular Rise of a Black Power Icon*. New York/London: New Press.

Scharf, Kathryn, C. Levkoe, and N. Saul. 2010. *In Every Community a Place for Food: The Role of the Community Food Centre in Building a Local, Sustainable, and Just Food System*. Toronto: George Cedric Metcalf Charitable Foundation.

Schlesinger, Arthur M. 1917. “The Uprising Against the East India Company.” *Political Science Quarterly*. 32 (1): 60–79.

Seale, Bobby. 1970. *Seize the Time: the Story of the Black Panther Party and Huey P. Newton*. London: Hutchinson.

Singh, Nikhil P. 1998. “The Black Panthers and the ‘Underdeveloped Country’ of the Left.” In Jones, *The Black Panther Party [Reconsidered]*. 57–105.

———. 2004. *Black is a Country : Race and the Unfinished Struggle for Democracy*. Cambridge, MA/London: Harvard University Press.

Steele, Anim. 2010. “Youth and Food Justice: Lessons from the Civil Rights Movement.” Oakland, CA: Food First/Institute for Food and Development Policy.

White, Miriam E. 1988. *The Black Panthers' Free Breakfast for Children Program*. Madison, WI: University of Wisconsin–Madison.

Zinn, Howard. 2003. *A People's History of the United States: 1492–Present*. 3rd ed. London: Pearson/Longman.

Chapter Eight

BEYOND VOTING WITH YOUR FORK: FROM ENLIGHTENED EATING TO MOVEMENT BUILDING

By Josh Viertel
Slow Food USA

WHEN I TALK to a crowd of people who are new to the food movement, I often begin by asking them, "How many of you have committed an agricultural act in the last 24 hours? Please raise your hand." In crowds of over 200 people, I usually see about six hands go up. I call on them: "What did you do?" "I watered my tomatoes." "I mowed my lawn." Occasionally, I'm surprised to hear that someone collected eggs from a backyard chicken coop, but most of the time, there are a few home gardeners in the audience, and that's it.

I then ask the question, "Well, how many of you have eaten in the last 24 hours? Please raise your hand." All hands go up. The message I'm trying to get across is simple: If you eat, you are involved in agriculture. "Eating," as Wendell Berry writes so eloquently in his essay "The Pleasures of Eating," "is an agricultural act."

Lately, more and more people have begun to get this message. I've seen radical shifts in the responses I get to my first question. First of all, more people are taking an active role in growing their own food, and they are doing it in unexpected ways. Recently, while I was on a college campus, in a lecture hall with around 350 students, a young man in the back row raised his hand and announced, for all to hear, "In my dorm room closet, I've got grow lights and a hydroponic operation." The room erupted in laughter as we, of course, imagined he was growing marijuana, and broadcasting that fact to 350 peers and his professor. It turned out he was growing cilantro.

Secondly, more people understand that, in eating, they are connected to agriculture. More and more people are responding to my first question with, "I ate lunch." Once, a young man in the front row raised his hand, clutching a can of Coca Cola, and responded, "I am drinking this Coke. It is sweetened with high-fructose corn syrup. That means that I am supporting a heavily subsidized, heavily polluting agricultural commodity, and I'm committing an agricultural act. Not a very good one though." He understood that behind his beverage was a story.

This has been a foundational idea for Slow Food: there is a story behind our food. And we believe it ought to be a story that makes us proud. Sadly, most of the food we eat has a story behind it that we would be ashamed to tell—like the Coke in the hand of that young man, though often worse. Our food consistently makes us sick, causing food-born illnesses like salmonella and E coli, and diet-related diseases like diabetes and hypertension. But even just *hearing the story behind it* can make us sick, too—sick with images of manure lagoons, sick with the sight of tortured animals, sick with the smell of thousands of miles of polluted waterways, and sick with accounts of labor abuses so severe, and working and living conditions so dire, that they are sometimes hard to believe.

But that isn't the only story. There is a different kind of food out there. It is good for the people who eat it. It is good for the people who grow it. And it is good for the planet. It has a story behind it that would make us proud. At Slow Food, we call it good, clean, and fair food. I have been supremely lucky to be able to grow and eat this food, and to help people make it a part of their lives.

People are beginning to understand that, as eaters, we are cofarmers. We create our food system by the choices we make about what we eat. More and more people feel that we should apply our values to the choices we make about food. Sure, everyone's values are different, but the truth is *anyone's* values will do. The problems with food and farming don't come from people holding the *wrong* values; they come from people not applying the values they hold. If most people ate according to their values, most of the problems caused by food and farming would go away, because no one's values can accommodate the status quo.

That notion, that we should eat food that reflects our values, has been at the forefront of Slow Food's thinking about how we should go about changing the food system. "Vote with your fork!" has become a battle cry of the food movement.

The idea that we should be proud of the story behind our food is built on the concept of empathy—the ability to imagine, and even feel, what it would be like to experience the conditions of another. I can imagine what it's like to work in a tomato field owned by an abusive grower, so I participate in a tomato boycott, even though I am not a tomato picker. Empathy reaches beyond people, and can be applied to animals and to place: I don't want to support chickens being debeaked, or hogs having their tails cut off, so I buy meat from a farmer who has practices by which I can abide. I don't want to support the chemical pollution of groundwater, so I buy organic, even though my well is elsewhere. This empathy extends to the farmer, too. I empathize with my friend Lynn, who sells cheese and yogurt to me every week at my farmers' market in Brooklyn. When I stop by on a cold, sleety day in February, I am glad to know that I am not only buying groceries; I am casting a vote for grass-based, sustainable, local dairy production. And further, I am not only casting that vote; I am letting Lynn know that I care about *her*, that I appreciate her braving the weather, that I count on her each week, and that she can count on me.

I see that sort of empathy as the foundation for a commitment to "vote with my fork." But just as empathy leads me to imagine the impacts of my daily actions on other environments and other people, empathy demands that I imagine the experience of people who have less choice than I do. Empathy calls solidarity into being. And I'm struck by the uncomfortable fact that not everyone can buy yogurt from Lynn. Not everyone can vote with his or her fork.

Can activists keep the Fairtrade market from undermining the movement that made it?

Christopher M. Bacon, Ph.D. Santa Clara University, CA

Fair trade is a strategy that smallholders, cooperatives, and food justice advocates can use to build collective power through the creation of a fairer and more environmentally sustainable market. From the 1940's through the 1980's, alternative trade organizations connected southern women's artisan groups and Mesoamerican smallholder coffee cooperatives with European and North American activists seeking to create a more democratic economy, effectively co-launching this movement as a North-South partnership.

Direct solidarity-oriented relationships initially moved small volumes with less quality requirements but provided a tangible alternative to the global commodities markets and the commercial free trade system. Fairtraders prioritized smallholder cooperatives contributing to their efforts to gain a foothold into a "different market" that shared risk, provided better prices and increased access to rural development projects. These advantages often increased smallholder's political economic power and improved their land tenure.

The more recent launch and aggressive promotion of a product-based Fairtrade label and certification has enabled larger corporate participation and facilitated a dramatic increase in sales—now topping 3 billion dollars. Unfortunately, the expansion of fair trade sales through conventional industrial outlets has resulted in another, disappointing, trend of selective enforcement, low returns to farmers and sliding standards, suggesting that fair-trade is not as fair as it used to be.

A split is growing within the fair trade movement as alternative trade advocates and many smallholder organizations become increasingly disillusioned with the market-centric mainstreaming strategy and the unaccountable governance structure among many Northern certification agencies. The voices of smallholders, alternative trader organizations, and activists are steadily being muted. Unless fair-trade can renovate its governance structure and re-root it its original, more transformative components, it risks becoming alienated from the growing progressive food movement.

Full Article at: http://www.foodmovementsunite.com/addenda/c-bacon.html

If dinner is a democratic election, and we seek to change our food system via our votes, we need to look squarely at the fact that, in many electoral districts and for too many people, there are no polling stations because there is only one candidate, the incumbent: fast food. And even if there were other candidates, it wouldn't matter, because many people can't afford to cast a vote for anyone but the incumbent.

I've said that if most people ate according to their values, most of the problems caused by food and farming would go away. I still believe that to be the case, but I am haunted by one simple and deeply troubling fact: so many people *cannot*. They have little money, they have little time, they may not know how to cook, and they have no access to ingredients, let alone ingredients whose story could make them proud. What they have access to is food that makes them sick, food that hurts the environment, and food that is grown and picked by people whose time and labor is degraded.

Lots of people *can* vote with their fork. And they should. If all who could did so, then things would get better. But I don't want to pretend that if everyone who *could* vote with their fork did, problems that stem from food and farming would be resolved. Too many people cannot, and it is neither productive nor fair to pretend otherwise.

For Slow Food, this calls our work into a different place. Our farmers' market tote bags won't get us into heaven anymore; there is more work to do. We can no longer simply say: "People should vote with their fork, and our work is to convince them of the value in doing so." We must instead say: "People who *can* vote with their fork should, and we should work to convince them of the value in doing so. At the same time, we need to right the structural injustice that prohibits many people from accessing or affording the food they need to keep themselves, their children, farmers, workers, and the environment all in good health."

This food we believe in—good, clean, fair food—can no longer be seen as a privilege. It must be seen as a universal right. Our mission is to create a world where everyone can exercise that right. We need to roll up our sleeves and work to ensure that the world we live in is one in which all people have the opportunity to cast a meaningful vote for a different food system.

This is new work for the Slow Food movement, but I believe that our movement—which began at the table, with a commitment to eating food that reflects our values—can bring power to the broader movement that is committed to justice. It can help to create a world where *everyone* can eat food that is good, clean, and fair.

All successful social movements have, at their core, the people who have the most to gain from making change happen, and who are most hurt by the status quo. There are always others, too: people who are driven by their values, by a sense of moral obligation, or by solidarity. But the beating heart of a movement will be the people who have something real to gain.

The food movement will ultimately be a soft and shallow thing if it does not have, at its core, the people who are most hurt by food and farming problems. We would shift the behavior of a percentage of people who have privilege, but we would not be the authors of a transformation. So we need, for both moral and practical reasons, to have the people who are disproportionately hurt at the core of the mainstream movement. For organizations that have power, that means sharing it, and redirecting it. It means making certain our work is first relevant to those with the least, before it is relevant to those with the most.

There is incredible beauty and opportunity in that shift. Imagine what the food movement would look like if its strength were *derived* from bridging divides in race and class—if shared food and shared work were a way of manifesting solidarity. Imagine what our cities and communities would look like if farmers' market organizations were always an integral part of the food-assistance program; if every public school had a garden; if growing and cooking food from scratch was an everyday practice that helped people save money, that helped people make money, that helped people to be healthy and happy. Imagine a world in which we can no longer point to the paradox of skyrocketing obesity and skyrocketing hunger as a symptom of injustice, and can instead point to the disappearance of both as a symptom of justice flourishing.

I am committed to the kind of movement that can make that world, and I believe the best way to begin is through building meaningful human relationships, through linking people and communities together around a sense of common purpose. Groups of people become communities by sharing work, sharing struggles, and sharing food. This leads to real, personal relationships; a sense of codependence and cocommitment. Once you've shared a meal with someone, or worked on a project together, you view each other differently. You're more likely to take care of each other, and, I believe, you're more likely to stand together and work for change together.

When it comes to changing the world, a big mailing list helps. But it isn't much good unless it is a list of people who care about the issues, and are willing to do something about it. People in a network need to be *engaged* in order for that network to have power. I've come to believe that a sense of spiritual uplifting and connectivity—the sort of connection you get from regularly sharing a meal together, working on a project together, or hauling in hay together—is a prerequisite for that engagement.

That connection can begin at a potluck, a volunteer workday, a canning workshop; while setting up a new farmers' market in a low-income neighborhood, or breaking ground on a new garden for a public school. It may directly better the world; it may not. Either way, it links people together, makes them feel like a part of something bigger than themselves and leaves them connected to a group of people who share a common purpose. Ultimately, that engagement is what allows us to change the world. Make no mistake; changing the world takes much more than getting people to attend potlucks. But human relationships are the foundation for social movements, and they are built out of shared experience, shared food, shared work, and shared struggle—breaking a sweat together and breaking bread together. Pleasure is a means as well as an end.

Millions of people have been inspired by Michelle Obama's garden at the White House, moved by the writings of Michael Pollan or Barbara Kingsolver, or angered by what they saw in the movie *Food, Inc.* Millions of people make an effort to shop at their farmers' market, to buy organic, or to cook food from scratch for their family. Millions more worry about what their kids eat at school, or are struggling to buy real food for their families—either because there is no place to buy ingredients, or there is no money with which to buy them. Each one of these people is frustrated, inspired, anxious, or pissed off, and each is primed to play a role in a larger social movement. Each is primed not just to complain over coffee with friends, not just to shop a certain way, but to truly become an agent of social change. Our role in the food movement is to help people make that leap, to help them in their transformation from a concerned individual to a force for change. I've seen people make that leap.

Leah DiBernardo in Temecula, California, wanted her daughter's school to have a garden, and for that garden to supply the cafeteria. She rolled up her sleeves, built relationships with the principal and with other parents, and brought that garden into existence. Leah's story inspired other parents at other schools in her community. As leader of the local Slow Food chapter in Temecula, Leah has now helped organize parents and teachers to create gardens at *24* schools in one small city.

On Labor Day in 2009, over 20,000 people—parents, teachers, farmers, college students, and citizens—got together in cities and towns in all 50 states and staged public potluck/demonstrations (we called them Eat-Ins) to demand that Congress pass legislation that helps schools serve healthier food for lunch. Most of the people who participated in that day of action already cooked and ate food that reflected their values, but they didn't necessarily identify themselves as advocates, or as agents of social change.

Participating in that day of action started to shift their orientation. During and after the potlucks, they gathered petition signatures, wrote letters to Congress, and made phone calls. A total of 160,000 people signed petitions or wrote their legislators.

In late November 2010, Congress passed the Healthy, Hunger-Free Kids Act. The bill represents the first noninflationary increase ever made in school food, and while there were many compromises along the way, the final product holds lunchrooms to higher nutrition standards, allocates funding to farm-to-school programs, and simplifies the bureaucratic process for low-income students seeking to register for free and reduced-fare lunch. In addition, through working to get a better bill, 160,000 people came together to push for something, and that in itself is a sort of victory.

The relationships built through those Eat-Ins may ultimately make for more significant and lasting change than legislation. Take Tarrytown, New York. The Eat-In that took place there brought together the predominantly white community of South Tarrytown and the predominantly Hispanic sections of North Tarrytown, because both wanted the same thing: better food for their kids.

The Eat-In had empanadas, a piñata, local legislators, even a member of Congress. Community leaders Gloria Sepin and Anna Lopez helped organize the event, translated materials into Spanish, and made certain their community was represented. In a meeting with the local Slow Food chapter, Gloria said, I want you to know how much it means to us that we were asked to be here. In all the years we've been living and working in Tarrytown, it is the first time the white community really invited us in to work together, and to be a part of a community-wide effort to make things better.

A seed was planted. Friendships were formed. Lines between the two communities have been blurred, and friends have continued to work together to raise money and bring in volunteers. Together, they built a beautiful garden at a housing project, then another at a community center. Versions of this story played out in hundreds of communities across the country. And they show that the food movement has an enormous opportunity to build connections between people, to better communities, and to build power that can yield structural changes at the national, or even international, level. It is an opportunity that exists every day, in the hearts of everyday people.

There is astounding unrealized potential power in the millions of people who have read an article by Michael Pollan and felt frustrated, or struggled knowing that their child is eating a terrible school lunch. What transforms that potential power into actual power is almost always a meaningful human interaction. People became active participants in Slow Food's school lunch campaign because they went to an Eat-In, were moved by what they saw, and were excited by the people they met. Parents and teachers in Temecula started those 24 school gardens because they got to know Leah, and she inspired them.

I want to be a part of a community as well as a movement and, because of the nature of food and the food movement, I believe it is possible—necessary, even—to be part of both. I can imagine the table giving people a source of pleasure, and also a sense of community, of connectivity, almost like the church does for some, and that out of that connectivity we can build power for change. I believe this will lead to a richer world for those who get more involved, and, ultimately, to a more engaged and effective movement.

It is vital to remember that pleasure and the table are not just ends in themselves, but organizing principles. We need to use what we know—the shared meal, the garden planted together—to drive for social change. When the United Farm Workers were banned from picketing, in their effort to gain farmworkers' rights in California, they worked with their Catholic constituents and the church to hold Mass services for the public at the entrance to big farms. They used the traditions, methods, and rituals familiar to them (and conveniently, the police would not shut down the service). They knew how to hold Mass; we know how to have dinner. We can use the power of the table to bring people together, to learn from each other, to express our shared values, and to build love and power between us that ultimately leaves us positioned to change the world.

Pleasure can be a radical force for good. Reaching beyond voting with our forks doesn't mean abandoning the farmers' market or the table, but seeing them as a source of strength in the context of a broader struggle in which we are all engaged.

Emma Goldman said, "If I can't dance, I don't want to be part of your revolution." The need for change is urgent, but we ought not to repress pleasure and celebration in the name of that urgency. Pleasure and celebration are a source of strength. Through dancing, or in our case, through sharing food and sharing work, we can make a better movement.

Chapter Nine

RACISM AND FOOD JUSTICE: THE CASE OF OAKLAND

Excerpts from interviews with Brahm Ahmadi,
People's Community Market.

MOST OF THE SOCIAL, environmental, and economic struggles in historically excluded communities stem from inequity and structural racism in our society. The 1940s through the 1960s was a period of time in America's history that was especially relevant to the current situation in urban, low-income neighborhoods. Food justice, in both the domestic and urban sense, is largely a response to political decisions and events of that time period that produced such tremendous disparity in food access and poor nutrition in the so-called "food deserts" that now exist in over 400 rural and urban communities in the United States. When we look back to this very important period of time we see development patterns, policies, and power relations that point to a fundamental problem in our social structure. The city of Oakland, California, is paradigmatic in this regard.

One of the most important factors in Oakland's development was the advent of the first transcontinental railroad, of which West Oakland was the final destination. Because of the transcontinental railroad, many people thought Oakland, rather than San Francisco, would become the main city of the Bay Area. However, the nature of transportation changed very radically over the next several decades, and that assumption turned out to be wrong. But many large factories and manufacturers saw great advantages to locating around this transcontinental railroad terminus. The railroad became a key driver of the industrial development in

Oakland's flatland neighborhoods. The Port of Oakland was also an important factor for industrial and population growth. This, along with the manufacturing industry related to the railroad, became the primary attractor for jobs and people seeking jobs in the area. The Second World War accelerated the rate of the region's industrialization.

In the 1940s, hundreds of thousands of people migrated to the region in search of work. Job seekers were ethnically diverse, but African Americans are most closely associated with this "great migration" to the Oakland area, as well as other port cities such as Richmond and Los Angeles. Additionally, as a result of the displacement of many San Franciscans who lost their homes and jobs in the great 1906 earthquake, Oakland's population increased rapidly in the early part of the 20th century and continued to grow throughout the middle part of that century.

Initially, there were not only lots of jobs for families chasing the American Dream (hoping to enter the middle class), but also significant entrepreneurial and small-business activity at the local level. Many people used their wage earnings from industrial jobs, combined sometimes with savings that they had brought with them, to start their own businesses. Perhaps the most famous such area at the time was the Seventh Street corridor in West Oakland, which was referred to by many as the Harlem of the West. It became a true economic and cultural center for the African American population, most of which migrated from the South. For a time there was also a strong presence of Polish and Italian immigrants in West Oakland, with economic centers of their own. Today's Chinatown was a fledgling economic center for the Chinese at that time as well.

All of these various trends in regional migration, industrialization, and economic development caused the industrial economy of Oakland and the Bay Area region to grow rapidly. For a brief period, there was a flow of economic activity connected to the wartime infusion of capital. But it did not last for very long.

Beginning in the 1950s, suburbanization would begin to pull the bottom out from under it.

The 1940s and '50s can really be credited as the period in which our modern urban planning ideology and doctrine first began to take shape, rooted in patterns of segregation, and keeping different working communities apart. During this period—which was still part of the Jim Crow era—very significant segregation activities occurred; work camps were developed and allocated for very specific racial groups—only Chinese here, only African Americans there, and so on. It didn't matter that they all worked in the same facility. This geographic "racialization" fed into the suburbanization that followed.

These development patterns, intended to facilitate suburbanization, were essentially used by Oakland's white political and economic establishment to dismantle and undermine nonwhite communities. As suburbs formed, so did the idea that they should be the spawning ground for the new middle-class American Dream. A home in the suburbs meant land and a single-family home in a peaceful community where appliances made every aspect of life comfortable and convenient. The entire structure of our economy had to shift in order to support this (created) growing demand for suburban living. The areas that had real potential for suburban development lay on the other side of the Oakland hills in Contra Costa and other parts of Alameda County. Planners, policy makers, public officials, private banking institutions, and the real estate industry began to collude to ensure that these suburbs would be white communities. Explicit and implicit methods were used to ensure that the new suburbs would not include people of color.

One such method was redlining, a common practice by which banking institutions essentially drew red lines around neighborhoods to indicate where they would not lend. More significant than the act of keeping capital from flowing into communities, however, was the dynamic of capital flowing out.

The departure of the white middle class to the new suburbs, often termed "white flight," resulted in a tremendous draining of capital from the inner city. This dissolution of wealth led to the unhinging of the economic foundation for these communities. Since the new suburbs were intended for white middle-class families living in inner-urban areas, there needed to be new transportation infrastructure to facilitate their migration. This required the construction of freeways and mass transit for white suburbanites who still had jobs in the inner Bay Area.

Thus began the construction of the Bay Area Rapid Transit system (BART), and all of the freeways Oaklanders now take for granted. These freeways were not initially created for urban dwellers. At the time, owning a vehicle was still a privilege reserved for high-income families; therefore, freeways were constructed with a particular class in mind. Deciding where to establish the BART line and the freeways also served as a means for various policy makers, home association groups, city councils, and other groups to dismantle the economic foundation of the communities they wanted to eliminate. The Seventh Street corridor in West Oakland, for example, was destroyed when the policy of "eminent domain" was used to seize most of the area for the construction of a BART line that passed right through the heart of the community. Although eminent domain is legally supposed to benefit everyone, it was used as a means of economically destroying African American communities. Similar developments like the Cypress Freeway and Interstate 880 followed.

The T-shaped economic center in the African American community in West Oakland, made up of the Seventh Street corridor and what is now Mandela Parkway, was destroyed by these development patterns. Many families lost their businesses and were poorly compensated for the seizure of their land by eminent domain. Many argue today that this was a blow from which the community has never economically recovered.

This occurred while World War II was winding down and jobs were beginning to disappear. There was still, however, a preexisting manufacturing, industrial base that had come for the transcontinental railroad and the convenience of port access.

The designers of the new suburbs needed a way to move those industrial jobs. They offered incentives, rebates, and cheap land to manufacturers, and launched a marketing campaign advertising beautiful landscapes where new factories could be built. They employed subtly racist language regarding the suburbs' "desirable (read: 'white') work force." The incentives coaxed many prominent companies and factories into relocating to the suburbs. A good example of this is the Safeway supermarket chain. Safeway's headquarters were originally in Oakland. But the company took advantage of rebates and available land and relocated to Pleasanton, where it's been ever since. When companies relocated, Oakland's inner city lost most of its jobs and a huge base of local capital. This economic "implosion" accelerated the ongoing departure of white middle-class families from the inner city.

However, people of color who remained in the inner city were barred from moving to the new suburbs. Many had the drive and savings to do so, but were directly and indirectly prevented from moving. Homeowner associations signed covenants with real estate and banking industries, essentially agreeing not to allow people of color to buy homes in suburbs. This practice continued for 25 to 30 years. In many ways, these policies are continued by other means today, especially in the banking industry.

Another simultaneous trend was a revolution of the grocery store business model. Prior to the 1950s, there were no big-box supermarkets. Even Safeway stores were quite small, relative to today's scale. But as supermarkets moved to the suburbs they began to buy large amounts of land, at rock-bottom prices, and build much larger stores. This radical transformation of grocery stores was consistent with the emerging economic theory of economies of scale, efficiency, consolidation, and centralization.

Crucial to these larger stores in the suburbs were the large parking lots they constructed for the new, car-oriented postwar society. The bigger stores enabled supermarkets to consolidate the spending power of much larger geographic ranges and to make far greater profits by having significantly fewer stores.

As supermarket chains started shutting down their smaller stores the urban food-retail sector disappeared, along with the rest of the economic foundation. Many elders in these communities today tell stories of how different everything was back then, how there were many genuine mom-and-pop grocery stores—not just pseudo-convenience/liquor stores, but actual locally-owned grocery stores—that carried fresh quality foods and were rooted in their neighborhoods with deep cultural significance. Grocery stores in those communities were often more than just places of business. Many of these small stores were used for communal meals, debates, meetings, block parties, barbecues, performances, and all kinds of social events.

But the majority of these smaller mom-and-pop stores could not survive under these stressed economic conditions, and they also began to disappear as spending power fell through the floor. Many of these stores were sold to other immigrant communities that were forming, especially within the Arab community. Most of these were converted into liquor stores in which some money could still be made through the sale of alcohol and tobacco.

So it becomes clear that the basic story behind the current conditions of food access is rooted in systemic patterns of structural racism in policy, urban development, migration, construction, and industry. It's no coincidence that over 400 communities devoid of grocery stores today are mostly communities of color. These same communities also tend to have the highest rate of chronic disease and other diet-related conditions like malnutrition. Understanding the historical context of the modern food system is important as we examine how organizations are currently addressing the vital need for access to fresh foods in urban neighborhoods, and the public health problems resulting from this lack of access. It raises many questions about what long-term strategies can be most effective in changing the modern food system's structure. Knowing how we reached this crisis is fundamental to figuring out where the food justice and food sovereignty movements need to go.

Corporate Solutions?

The food movement's solutions to the food crisis must focus on the same issues that created it, and tackle them from all angles. Policy is an obvious one. We must also revolutionize our perception of land and the role it plays in society; we must examine land ownership, land use, and land tenure in urban neighborhoods. This has a direct correlation to environment and food.

Some food justice activists are increasingly focusing on developing entrepreneurial solutions for enabling fresh food access. But, given the ideological and structural underpinnings that drive the financial sector, it is proving to be extremely challenging to finance these social ventures. Despite recent buzz around the rising economic potential of inner-city neighborhoods, the local business community still isn't taking the idea of investing in food deserts seriously.

One new development that may help overcome the challenge of financing food retail in inner-city food deserts is increased sources of public financing. The Obama administration is giving attention to the food crisis and is preparing to launch a healthy food financing initiative. First Lady Michelle Obama is also getting involved through her Let's Move campaign. The United States Department of Agriculture (USDA) is showing interest, and the Department of Health and Human Services (HHS) is unveiling plans as well. Obesity and urban food deserts have become hot topics at the federal level, and funding is being rolled out. As a result, the corporate retail industry that previously had no interest in inner cities is now practically salivating over the public money being allocated to deal with food deserts. (Corporations could easily finance opening new stores in inner-city neighborhoods with their own funds, but why should they when the government is perfectly happy to hand them the cash?)

Imagine if the national answer to the food crisis took the form of a huge, publicly financed flood of corporations like Walmart and Tesco opening up stores in inner-city neighborhoods, using the exact same economic model they're using now. We could expect low wages, the destruction of small businesses and local economies, and all of the awful labor and supply-chain practices we're familiar with. We all know the model under which these large corporations operate, and there is no reason why they won't replicate the same essential business model in neighborhoods that need good, living-wage jobs that pay meaningful earnings and teach meaningful skill sets. The supply chains will also emulate the same model—behaving destructively toward the communities where they source their food products. So poor urban communities will also see their economies tied to the wealth and resource extraction from rural communities, with the usual negative consequences for local economies and the environment.

It would be ironic if corporate America, which helped create food deserts, took the lead in rescuing food deserts and food-deprived communities, yet this is a potential reality. When Michelle Obama launched her Let's Move campaign, she made a statement that on the surface was uplifting, but under careful analysis, is actually terrifying: she said that she wanted to see food deserts eradicated within seven years (a time frame that coincides with a two-term Obama administration, of course). We should be fearful because there is really only one group of actors that can make such changes so rapidly: large corporations. But many people understand that re-mending the torn fabrics of local economic systems cannot be done in such a rapid manner or through corporate models that are designed to extract wealth from local economies in order to maximize profits for shareholders. Nonetheless, most policy makers, community-development corporations, or various intermediaries around the country are not applying rigorous analysis or qualitative criteria to the question of which kinds of stores are best for food-desert neighborhoods.

For example, in early 2010 I spoke to a woman at the chamber of commerce in Oakland who said, "We're so excited. We're talking to Costco about opening a store on Mandela Parkway in an old steel factory. We really want to bring the big-box stores to Oakland. We see them in Emeryville, we see them in other places, we want them here." But one must question how these retail industries will rebuild our local economies, given that they are designed to *extract* wealth from our neighborhoods and distribute that wealth to distant and anonymous shareholders that place no value on the well-being of our communities.

This is why the food justice movement is in a battle to prove that there is another way: that we don't have to sell our local wealth, our land, our environment, and our health to corporate America just to bring some superficial change in an expedient manner. Unfortunately, expediency is a treasured value in our society, as our economy operates according to short-term, quarterly cycles. The same holds true of our political system. Michelle Obama's statement about eradicating food deserts within seven years reflects the short-term mode of political thought that is centered on election cycles. It's even tougher at the local level to get public officials to think in longer terms, when they're always looking for ways to position themselves for their next campaign. Try telling them, "We have this community-based venture that is a fundamentally redesigned business model intended to address the various barriers to operating and developing grocery stores. It's intended to bring back the social fabric of what stores used to be like, not just to offer some good public relations around making fresh food available but to really engage in improving health awareness and education, and promoting lifestyle changes. But it's going to take a long time to develop, to finance, to launch, to make profitable, and to see the changes we're talking about." I've yet to meet a public official who can get behind that kind of a time frame and process, simply because it wouldn't happen by the time the next election came around (or even the one after that). Therefore, public officials continue to look to the big-box retail industry because it is expedient and, with a combination of private and public financing, can set up shop and be operating in time to make a great public splash for the elected officials who support them.

Food Sovereignty from the Ground Up

Only committed grassroots organizations can truly help our communities. We have the capacity to imagine something better than the amoral system that corporations are designed to impose. We have the power to dream. But if we want to solve the food crisis, we need to begin turning our dreams into practical visions for overcoming the current system. Most of the solutions food movements are currently proposing would take 5 to 15 years to fully take effect, which is too slow for most donors, philanthropists, and banks to support. These proposals often need much longer payout horizons than investors and banks are accustomed to. Like companies and politicians, financiers are also stuck in a short-term paradigm. As much as many of them might like to see such ventures become a reality and succeed, they are trapped in a mentality that demands quick returns on their investments. At the end of the day, the system of capital is the giant brick wall the food movement keeps running up against. The issue is not simply a lack of capital but capital itself, and the present ideology behind it.

Society's modern form of capitalism is the fundamental barrier that has gotten us into this mess and is keeping us in this mess. So it's critical that we change the nature of capitalism in order to get ourselves out of this mess. We all live in a capitalist society, and we obviously need capital in order to operate. Some food movements have come up with remarkably creative means of getting things done on tiny budgets, or even on no budget at all. But we still have yet to see those movements provide something of equal stature to big industry and corporate America. Most Americans are ready to get behind an alternative food system. But in order for that to happen, it has to be as convenient and prevalent as the current one. The food sovereignty movement has not yet reached that stage. We're not at a level where our practices—community food systems, local food systems, food justice work, food enterprise work, urban agriculture—are equal to the industrial options we seek to replace.

There is also an issue with how most nonprofits are structured. Most of today's nongovernmental organizations are very grant dependent for their work. And while their activity often does generate income, by and large they're simply not structurally designed to grow large enough to end their dependence on outside funding.

The last 10 or 15 years of the food sovereignty movement could be characterized as its formative years. It's important to think about the next 10 years, what the next phase of the movement should be, and what ultimate destination it's trying to reach. The movement is surrounded by excitement due to a growing trend in local foods and food ethics. But how much faith can sensibly be placed in a trend? History indicates that trends rarely last, especially when they're consumer or media driven. It may be years before the local-food trend flattens out, but it may eventually fade, and then food movements will be left with the real question: What do we have that can stand on its own, without dynamic consumer and media support? An answer is imperative, because corporate America is already preparing to highjack the new popularity of local food and present itself as the only hope for food deserts and impoverished communities.

The food movement is still at a largely experimental stage; we're still trying to figure out what works. We've seen an exciting level of effort and experimentation that runs the gamut from urban agricultural projects to various kinds of food-distribution activities. But we don't yet know what can effect lasting change. It's still not clear how the food movement can help communities economically. Our projects may be able to make a significant contribution to nutrition. But if they can't also make an economic contribution, how can communities sustain their advances?

That is why unrelenting activism and social entrepreneurship are essential to our success. We have to identify clear and specific opportunities for action in people's consumerism and in use of the media. The hype around food movements today is essentially mass-media driven, and while it's exciting, the question remains as to how we can pull that vibrancy down from the Internet ether into town-hall meetings and public rallies. We need to ride this tremendous momentum and create direct-action opportunities for the public. The vast majority of people that are passionate about these issues still don't see many opportunities for involvement.

We need to see greater turnout in public demonstration. We've witnessed the proliferation of food sovereignty websites, blogs, and recipe sites; we've seen a huge proliferation of farmers' markets, community supported agriculture (CSA), and different types of direct-marketing solutions. But we're not seeing enough political activism in the communities with the most at stake in the food movement. Only a proliferation of activism can give the food movement the momentum it needs. Policy change through a prescriptive approach is a very slow process. The Oakland Food Policy Council is an example. While it has some fantastic ideas, how long will it take for those new policies to be institutionalized and begin to break down structural barriers? It could be years, or decades. We can't afford to wait that long.

People are suffering right now from extreme health issues. West Oakland has a 48% rate of obesity, or unhealthy weight—almost half the community—and 67% of West Oaklanders have diabetes, which is two to three times the national average. That has a tremendous economic impact. If the main family provider falls seriously ill and is unable to work, it results in a significant income loss for the family—even with potential disability support. Even more important is the cost of health care. In 2009, the average health care cost per capita was measured at $7,600. In low-income neighborhoods with two to three times the rate of incidence of disease, costs are two to three times higher. This is a huge drain on a community's dwindling economic power. In addition, there is a pattern of dollar leakage—specifically food-dollar leakage—in neighborhoods with no food outlets available within their boundaries. Residents have very little choice for local food shopping, other than what is available at the corner convenience or liquor store. When you shop outside your neighborhood, you are giving your share of your community's money to an outside location. This may have a positive multiplier effect for that district, but certainly not for the one that you live in.

Rebuilding Political Power

Attrition of community economic power results, in part, from the outflow of spending as a result of "under-retailing" within local boundaries. This means lost job opportunities, lost opportunities for the multiplier effect of spending in the local economy, and lost tax revenues needed to fix the streets and support the schools. It is no coincidence that McClymonds High School in West Oakland is the third-worst performing high school in California, or that it has the lowest graduation rate in the state; there's an evident correlation between food security and academic success.

We have to see communities like West Oakland lead the activist charge in the food movement because, first, they know best the severity of the situation, and, second, they want to determine how the food system affects their neighborhoods. They have the potential. Before suburban society declared war on them, communities like West Oakland and many others back in the 1940s were sustaining themselves primarily on their own entrepreneurial activity. In the case of West Oakland, had the BART line not been dropped on Seventh Street, and had the Cypress Freeway not destroyed a black-owned small-business sector, we would see active main-street corridors remaining vibrant in the face of the manufacturing-job collapse.

Still, activism in these communities is a challenge since so many people are accustomed to being rejected, ignored, or worse. West Oakland is where the Black Panthers; the ones who invented school breakfast; came from. tFirst they were ignored, then ridiculed, then harassed out of existence. Learning from the experience of the Black Panthers, we must realize that the modern food movement must be more than just a small group of disenfranchised people trying to foment and agitate. And it's only through a broader political picture—mass public turnout—that such a movement can expand. Most elected officials don't see disenfranchised people as tremendously relevant in terms of their campaigning, so they need an incentive to fight for them. Unfortunately, voter turnout is very low in neighborhoods like West Oakland.

Those most impacted by problems should be the leaders in the fight against them. They should be at the forefront of generating ideas and solutions. This issue is also important as it pertains to the role those from a privileged background can play: supporting and forming bonds of solidarity with impoverished communities. Right now, tension exists not only between people of different backgrounds but between movements as well. I see a great divide between food justice networks with communities of color, labor movements, and farmworkers on one side; and the sustainable, local-food, gastronomic-cultural movements on the other. There are tremendous disconnects between those groups, as can be seen in their rates of growth and change. On the one hand, we see tremendous proliferation in CSAs and farmers' markets, but on the other, we haven't seen significant change in food deserts or the labor conditions of food and farmworkers. That disparity of progress reveals the food movement's internal problems. We need large-scale, multicultural movements. But such movements can only work when we all begin to accept and live each other's values. Our current concept of leadership has to be corrected. Those of us that are most naturally drawn to it or able to take it on easily really need to reconsider our roles. That doesn't mean we don't have leadership to offer, but rather that it should not necessarily conform to the Western, patriarchal image of what leadership looks like—where leaders become the symbols of their efforts. That concept has to change, and in that change we may start to see what democracy really looks like at the grassroots in this great experiment that is our food movement.

Chapter Ten

CONSCIOUSNESS + COMMITMENT = CHANGE

A Conversation with Lucas Benítez of the Coalition of Immokalee Workers

I THINK THAT WE need to create alliances between all the food movements. We all want a healthy food supply and everyone in the chain to be treated with dignity—from the production worker to the consumer. But we are facing a monster: the corporate world. They are only interested in money and profits. We have to be clear, relentless, and determined to do what it takes in our communities to create change where we want it. Eventually these corporations, if we hit them where it hurts—if we go for their profits—will be forced to change the way they do business. Maybe we can't make Walmart disappear, but we can change the way they do business. The power is in our hands. The first thing we must to do is develop consciousness and commitment to create change.

We founded the Coalition of Immokalee Workers in 1993 because of the situation that workers in Immokalee were living in, and still live in. In 1995, we went on our first strike. We focused on our bosses, on the contractor, and on our direct superiors because at the time, we felt that they were the problem.

After a closer analysis, we realized that the rancher and the contractor are only a small branch of a much larger tree. We have always said that you can always prune the branches, but the tree will grow back. The coalition continued having marches and strikes, but we saw that we were not going to change things that way. If we don't pull up the tree's roots or water it with different water, we won't get change. The tree cannot drink polluted water; it needs fresh water if it's going to grow back with good branches and good fruit. We have slowly been creating change. We have switched the water to better water so that it bears better fruit. It's still contaminated, but it is getting better and getting cleaner. For example, there are no more cases of physical abuse; we have reduced the instances of wage theft; and we saw a small increase in salaries. We haven't gotten 100% of what we wanted, but a small change is still a change. We realized that to change the agricultural tree, we had to target the big corporations that have a strong influence in the agricultural industry in this country.

The first year we were able to get over $100,000 back in withheld wages for our coworkers. This reduced the back wages down to less than 20% each year. It's still a problem, but on a smaller scale. Physical abuse was a frequent problem in the fields and this has decreased. In those days, we had three or four cases of abuse per harvest season. So we marched to our boss's house and picketed in front, like a boycott against him, although at that time we did not see it as a boycott. We just said, "We will no longer work with this contractor because he hit a worker and we will not accept that." Everyone stopped working for him and the rest of the contractors took notice. We have not had any physical abuse reports since 1996.

These are big changes, but at same time, these abuses are still happening in other areas. Still, we have set an important precedent.

The Agroindustry

Corporations are changing how business is done, changing our communities. If we want to change that, we have to change them. Many people do not know that in this country family farms no longer produce all of our food. In the past, local family farmers were the main producers; they would sell their goods at the local markets. This is not the case today. Today, corporations have changed both the face of agricultural production and the market. This has happened right here in Florida. A local store here in town that had been selling fresh produce for years was forced to close its doors because Walmart had opened a store nearby.

On a daily basis, farmers' families are being ruined within the greater system. Big corporations put tremendous pressure on small family farms. They don't want to buy from 200 family farmers; they prefer to buy from three industrial farms. They don't care how the product is produced, and they don't care how far they have to go to sell it, whether it's New York or Washington. The small family farm hires many workers. When a small farm gets put out of business, they are also killing off jobs for hundreds of workers that had a relationship with their boss, jobs that were better paid and more humane. That relationship no longer exists. That rancher moved on to become a worker on his own land run by big agroindustry. Food sovereignty cannot be maintained that way.

We must develop consciousness and commitment to create change. That cannot be left in the hands of the government. A clear example was seen in Mexico. During President Vicente Fox's administration he said, "We are going to make it so that the small producer is the exporter of his product. If the world's avocados are produced in Michoacán then we are going to take them away from the bulk buyers." That sounded great. The farmers were happy, the producers of avocados were happy. What happened? Now the ones exporting the avocados are a company owned by Fox's family. They are the ones monopolizing all that small farmers produce. They take advantage of the small farmers and buy their produce for a fraction of the price; then they go out and sell it. In truth, Fox's family is monopolizing and being opportunistic. When he was president, he did everything so that his family could benefit. That is why, when we talk about food sovereignty, it is the community that must be involved, be the monitor, and be the one that is watching and making sure that they abide by the rules. Taking a government to trial in an international court takes a long time. A lot of time goes by before they reach a verdict, and meanwhile the people have to eat junk.

We have to start from the roots, with the community, so that the community knows what we are talking about. We understand our world, but if we don't work from the ground up we won't advance. It's like talking about a penny raise only amongst the Coalition. We understand, but if we don't get the community involved, they won't understand. They are the ones that will get the penny raise

The NAFTA Flu

By David Bacon

Bacon's analysis of the intersection of immigration and labor in the food system draws on the example of Mexican migrants working for Smithfield Food's hog-raising and slaughtering facilities. While outlining the injustices of working for Smithfield (no worker protection; environmental damage; severe health repercussions), he connects food injustice and insecurity as manifestations of the lack of labor rights.

> We can't have food sovereignty that looks at the food needs of people as migrants that doesn't also see that the reason why people [experience] food insecurity is because they don't have legal status or people are being treated as . . . a very exploitable labor force . . . where the main objective of the system is to ensure that people work for the lowest wages as possible meaning that people's income is so low that people can't even guarantee themselves the food that they need to eat much less send money home to their home communities where people now depend on those remittances in order to be able to feed themselves.

Bacon discusses the importance of political reform as a means to address injustices within the immigration system, and food insecurity for migrant populations and their families.

> An immigration reform that included a jobs program for communities with high unemployment would put a floor under the income of working families, while removing the fear of job competition. We need a system that produces security, not insecurity. Major changes in immigration policy are not possible without fighting at the same time for these basic needs. But these are needs that working people have in common, not just immigrants. By fighting together, people can create a more just society for everyone, immigrant and nonimmigrant alike.

Full Article at: http://www.foodmovementsunite.com/addenda/d-bacon.html

and they will see that as a gain. If it is later taken away, they will feel as though that benefit was taken back. They are not aware. The same goes for the government; they give you something for a while then they take it. And people say, "We ate well for a while and then it was over." If we educate our communities, then things will be different. Our mathematic equation is C + C = C. That means: Consciousness (awareness) plus Commitment equals Change. That is what we do. If we don't have those variables, we can't create change.

It is very important that young people become involved in this movement, because they have to make the change from compulsive buying to responsible buying. This is the first thing that must be done. I think young people are doing this more and more, especially those that are focused on eating well—vegans, vegetarians, and others. What happens is that sometimes they don't look to see what is behind the production of their food. They pay more for good food, but this is not passed on to the workers. They need to get out of that box where they have the luxury to be studying, the luxury of having a job without having to think about where their food came from. Young people in universities have power. We have to start there. They live there, they eat there, and that administration feeds them. The students pay for that food. Because they are the university's clients, they can change the way the university does business. Many universities offer fair trade coffee because students asked for it. The students made that change. The university changed because their clients asked them to. Starting with their own universities, young students can have a big impact.

Most important is mutual respect for each other's struggle. Many times, there are academics, people that believe they are industry experts in a certain area, but they don't have experience working on the ground. These people have conducted studies, and they have spoken to workers, but they have not worked in the fields. It's nice that they write it up and that their voice is heard, but just the same, they have to recognize that the ones *working* are the experts, and they must give them recognition for that. A mutual respect amongst everyone is what can lead us to a world with sustainable food.

Poor people are the most marginalized, and we consume the worst food in the food chain. What we get here is what they don't want at the nicer stores where they pay more. We have the experience of knowing how to work the land; we can share our experiences in cooperatives. The world of co-ops is growing more; people are creating their own community gardens. This is important knowledge because there are people that have grown up in places like New York City that have no idea where the tomato in their salad came from. This is a true story: I have a friend from New York City that asked me how it was possible for cucumbers to grow in vinegar. She had no clue! There are many young people out there like that; it is their reality. We have to connect the urban world, the city, with us farmers; not only is this interesting, but also necessary.

We realize that it is a long road and that we have just begun. Other movements like the animal rights movement and the movements to save the environment have been on this road much longer than us; workers' rights have never been addressed like this. I think we play an important role as a coalition in the new socially responsible world. The consumers are also beginning to become involved. Consumers pay more for organic produce and for free-range animal products.

Raising Consciousness

Corporations know that they have to change to become socially responsible. The words "*socially responsible*" were not part of the vocabulary in the corporate world two decades ago. The words have become a fad, and corporations use them as makeup so that they look good in the media, while in reality, they continue having an ugly face. We want that pretty face to be pretty without makeup, we want surgery that truly changes it and leaves the face like new. We want them to actually be completely socially responsible. At the coalition we said to ourselves, 'if we are talking about social responsibility, we are the ones missing in this picture.' Before, they never spoke about the workers. With the Fair Food campaign, we have started to enter the picture where the worker traditionally did not have a place—at the Kellogg, Slow Food, and Bioneers conferences that we have been invited to people say, 'We are seeing things that we had not thought of before.' What good is it to be vegan and eat a plate of organic mushrooms, if the salaries paid to the workers are lousy, and they have no medical benefits and no overtime pay? This means that nothing changes, and they are just paying more for organic mushrooms. That is how we come into play with the existing food movements, and how we make them stronger.

We are all connected; we are all in the same basket—from the office worker to the farmworker. We are connected because if I don't harvest the vegetables, you don't eat. If the butcher does not do a good job, you don't get a good steak. It is all a chain. Some produce, others consume; at the end of the day it all ends up on a table. Generally, we don't consider how the food got to our table. It is important to take a moment to reflect: 'Is what I am doing right? Can I do more?'

Pay attention, and respect the people in the food chain, because a New York waiter could be exploited, and you indifferently eat at the restaurant. One must open one's eyes, one's mind, and one's heart to make the food system change, to make it healthy. Your plate of food may look pretty, but it is filled with the tears and sweat of exploitation. The tomato that makes it to Whole Foods will never make it to the Winn-Dixie here in Immokalee; it's not the same quality tomato. Here we only get the cheapest tomato, and that is what happens in the poor neighborhoods.

Even though there have been some changes, the majority of the worst abuses are still happening. We have to develop different strategies. You have to set a precedent like the one we set with McDonald's to create change in other corporations. We have been able to get corporations to accept responsibility for farmworkers who make up the bottom of the production chain. If we were able to get them to accept responsibility for us, other people in other industries in other communities can demand the same from corporations. The precedent we set gives them a solid platform to make their demands to the corporations. We are not going to tell people what to do in their communities, but we have created a platform so that other communities can do something.

Our current strategy is to change people's attitudes and the way business is done. We are not saying we are going to boycott tomatoes in Florida. We don't believe that is a good strategy because we need to work, our families need to eat, and American families need tomatoes in their kitchens. That is why we do it corporation by corporation. Tomatoes are not like apples, peaches, or grapes—products that you can choose to eat or not to eat. No, tomatoes are an important ingredient in lots of foods. That's why we're doing it corporation by corporation. We attacked Taco Bell, but people had options to go to other fast-food restaurants. Once we got Taco Bell to the negotiation table, it was no longer necessary to boycott McDonald's or any of the other seven corporations. If we had to boycott them, we would have been closing one door and opening another, but it wasn't necessary. One by one, little by little, we are changing them.

Alliances

There are two important branches in this movement: the student branch and the religious branch. The young people played an important role in the civil rights movement—blacks, whites, Latinos, everybody; they wanted to change the situation. After the civil rights movement, the student movement fell apart. The students focused on their studies and they graduated. Today, we have big professionals and big brands without a mentality of change, without a social vision.

Students Forever was born out of the fair food movement and has gained force in the last few years; it is an alliance between farmworkers and students. Currently, there are student groups involved in the project throughout the country. This project is a wake-up call for them, not only to help us with the coalition and the tomato campaign in Immokalee, but also to create change in their own communities.

It is the same thing with the religious branch. For a while they were becoming more and more conservative: "We have to go pray and have God fix everything." Now, many religious leaders have told us that the fair food campaign "gives us the ability to practice the word of God, in our lives today; to change the situation for thousands of people." We have created alliances with Presbyterian, Unitarian, Catholic, and Evangelical churches. In our rallies, you can see Jews, Catholics, and Muslims coming together for a greater cause. An example is the archbishop of the dioceses in Orlando, Florida, speaking with pastors from other churches about how they can advance the campaign against Publix; how they can work together with their congregations to exercise some power and change the executives' minds at Publix.

We have worked very closely with other workers in the fast-food industry, supermarkets, meat processing, and with other workers in similar situations. We shared experiences with the leaders of the Smithfield campaign. Many of the Immokalee workers work in those packing plants. Contractors come here and take workers to work in the chicken industry in Iowa and other places. Many times, our members working in other places call us and ask us how to act when they have a problem. We instruct them to do the same thing we do here: they have to form a committee and speak with the supervisor to try and change the situation on the inside. Some of our members have participated in starting unions or in the collection of signatures to form unions in the packing companies.

We were able to establish a zero-tolerance policy for slavery, and we gained a penny in pay for the workers' wages; not only that, but workers are guaranteed to play an integral part in the monitoring and design of the conduct code. These are big changes, but at the same time, they are small because it is only true for the Immokalee community. We have not yet played a significant role in the food sovereignty movement, but I don't see any obstacles. We can be part of the movement if each sector is respected and self-coordinated. People played an important role in the civil rights movement—blacks, whites, Latinos, everybody. People like writers and industry experts that fight for fairness in the agricultural industry can help the cause tremendously by helping our voices, the voices that represent food sovereignty and fair food, be heard in the higher-up places. United, with each different experience, we are stronger.

Chapter Eleven

THE RESTAURANT OPPORTUNITIES CENTER

By José Oliva,
Restaurant Opportunities Centers United

AT AGE THIRTEEN, I came to the United States with my parents after being exiled from my home country, Guatemala. Food has always been at the center of my story and my people's story.

My family's exile was a direct result of the powerful relationship of United Fruit Company (UFC; now known as United Brands) with the US State Department. Back in 1954, the CIA sponsored a coup to depose Jacobo Árbenz Guzmán, Guatemala's democratically elected president. This was largely at the UFC's request after Árbenz nationalized lands belonging to them and redistributed them among Guatemala's landless peasants. The coup triggered a civil war that lasted 36 years and killed more than 200,000 people. My grandfather, Mario Gonzalez Orellana, was squarely on one side of that conflict, having served as vice minister of economics for Árbenz. My mother had been raised with clear pro-democracy, antihunger values; I was born with that fire in my belly as well.

We were forced to flee in 1985, when my parents began organizing anti-dictatorship student and labor protests. I remember watching bodies being pulled from the river that roared past my grandparents' house in Xela, my hometown. I was to leave all of the death and hunger behind . . . or so I thought.

We went to the Canadian Embassy in Guatemala City and begged them to give us asylum. But they refused, saying they had a quota and it had already been filled. They directed us to the American

Embassy, saying, "Don't ask for asylum. You'll never get it. Instead, say you're a teacher and you own a home. Apply for a tourist visa. Tell them you want to see Disney World." That is exactly what we did. Being the eldest of two boys, I pretended to be excited to go see Mickey Mouse, and we got the visa. The next day, we were on a plane to Orlando.

The change in setting could not have been more stark. Instead of goons with guns, there were old ladies with shopping bags full of plastic trinkets—useless, but shiny. Instead of dead bodies being pulled out of rivers, there were miles and miles of restaurants and shopping malls. Instead of hungry people protesting a brutal military regime, there were happy, sunny sidewalks filled with people laughing and joking. This is the way life ought to be, I thought.

After several weeks in Orlando, my parents made a decision to move the family to Chicago, in part because of a very hostile, right-wing environment in South Florida. Soon after we arrived there, my mother found a job at a fancy Italian restaurant called Giordano's, in one of Chicago's wealthy northern suburbs. I used to look inside through the restaurant's front window and think, "This is the job I want when I grow up." It was fancy and clean inside, and everyone was always smiling. My mother didn't work there very long. She left one day and didn't go back. It wasn't until much later that she told me, "*Hijo*, if you only knew how they treated women in that place. It was like rubbing salt in a wound." She was working an average of 60 hours per week with no overtime pay, and was constantly harassed, along with most of the other women in the restaurant. She earned minimum wage doing salad prep. My father was having trouble finding work, so what she brought home was the total income for our family of four. I would find out for myself years later what the restaurant industry was truly like for immigrants and people of color.

My grandfather always thanked the Creator for tragedies, no matter how devastating. In the *Popol Vuh*—the Mayan Bible—destruction is a form of creation. So, my grandfather accepted pain and suffering as a necessity for change. The coup that overthrew his boss's government was no exception. It brought him a peaceful life in the country away from the 11 consecutive dictatorships that would follow. When we came to the United States, in exile and shame, I tried to remind myself of my grandfather's peace and wisdom in his exile.

Because of my parents' sacrifice and many lucky breaks, I graduated from high school and went on to college. Since we had come to the US on a tourist visa we'd overstayed, I was completely undocumented at the time. I could not apply for student loans, grants, or even scholarships. I had to pay for my classes and books out of my own pocket. I did what anyone else in my position would do: I got a job. I found my first restaurant job when I was in high school, at a fast-food restaurant across the street from my house. My first day of work, I had to put on a giant chicken costume and walk around the block handing out flyers. My friends all recognized me (the suit had a huge open beak through which my face was plainly visible) and would walk behind me, kicking the suit's tail feathers. Apart from the brief humiliation, the job was hard. When I first burned myself in the kitchen (a second-degree burn) while dumping grease from a frying container, the manager shrugged and said, "It happens to all of us." He lifted his shirt sleeve to reveal several scars. During and after college, I worked at several other restaurants—Shoney's, Red Lobster, Francesca's—each considered "better" than the last. But for the workers, with the exception of a few "front of house" folks, the conditions were the same: minimum wages, no breaks, slips, falls, and burns. All I could think while working these jobs was, "There are more of us workers than managers. They can't possibly do all the work themselves. Why do we take so much abuse?"

ROC: Restaurant Opportunity Centers

On September 11, 2001, I was eating breakfast with my mother, watching the morning news, when there was an interruption in the broadcast, saying an airplane had crashed into the World Trade Center. The story that was to unfold was no exception to my grandfather's rule. The tragic events of that day led to the creation of a new actor in the restaurant industry: the Restaurant Opportunities Center of New York (ROC-NY). ROC would revolutionize both worker organizing *and* the restaurant industry.

Survivors from Windows on the World, the fine-dining restaurant at World Trade Center's top floor, had banded together to create an organization with a groundbreaking model to change conditions in the industry. The restaurant workers from Windows hailed from all over the world, an intentional choice by the owner, who wanted a staff that spoke every possible language so that tourists there always had someone to converse with. The staff were like family, and all made a good living. ROC cofounder and former Windows worker Fekkak Mamdouh says that within one year of working at Windows, he was earning $50,000—more than he'd ever earned before. He would also meet with several other Muslim coworkers for daily prayers in the World Trade Center's North Tower stairway.

Seventy-three Windows workers perished on September 11, 2001. The other 300 were left jobless. But they didn't sit around and mope. Under the leadership of Mamdouh and Saru Jayaraman, who had started several other successful local and national organizations, they came up with a plan for changing the very nature of the industry. They knew it wasn't enough to find new jobs. Bad jobs dominated the industry; even if a few former Windows workers managed to find good jobs, the majority would be placed in low-wage, no-benefits positions. They knew they had to change the industry's overall conditions.

They came up with a three-pronged strategy to create a new business model in the restaurant industry, one that ROC still uses today. Instead of the old "race to the bottom" model, it would lead to a new vision of "shared prosperity." The model's three prongs—research and policy, workplace justice, and promotion of the high road—are interdependent and intended to work as a synergetic wheel, energizing each other while engaging restaurant workers to create sustainability in their own industries.

Since its creation in 2002, ROC-NY has won nearly a dozen workplace justice victories against large fine-dining empires in the city. It has also developed groundbreaking research and a high-road association composed of several employers that provide benefits above and beyond what the law requires. This model's success in New York led the founders of ROC-NY to create ROC United in 2008. ROC United is a national organization composed of local ROCs in seven locations: Chicago, Los Angeles, New York, Miami, New Orleans, Detroit, and Washington, DC. All use the three-pronged model to improve conditions in their local restaurant industries.

A "dog-eat-dog" business

As ROC's founders knew well, the bad experiences my mother and I had had in the restaurant industry were far from isolated incidents. They were part of an entire business model. Workers call it the "dog-eat-dog" way of doing business. The key components of this business model are:

- **Contingent labor** As a worker, there is little consistency in work scheduling. You may be on a schedule one week and off the next, never to be called back—not fired, just not on the schedule. You may have 60 hours of work one week and three the next, depending on the whims of managers, the ups and downs of business, or other factors over which workers have no control.
- **Informality** Workers are often paid cash under the table, with no taxes withheld. No employee manual is ever provided. No rules, principles, or workplace policies are set; workers are kept uninformed and uncertain about their own rights.
- **A culture of legal violations** Labor and employment law is routinely violated, and rights are systematically denied as a way of creating a culture of inevitability and futility.
- **A classic divide-and-conquer approach** Employers hire immigrant workers and people of color for "back of house" jobs—as cooks, preps, and dishwashers, while hiring mostly

native-born white workers for the "front of house" jobs such as bartender and waiter. Besides creating a huge wage gap, in which people of color make an average of $4 less per hour than their white counterparts, this tactic also pits immigrants against native-born workers, and white workers against workers of color.

These four employment practices are not new. What is new is their simultaneous deployment, not in the small mom-and-pop diners, but the trend-setting, fine-dining segment of the restaurant industry—which is mimicked by the rest of the industry. These employment practices are now finding their way to many other industries and sectors, in essence "restaurantizing" the workplace and creating a new rat race to the bottom for all workers. Moreover, in an economic setting in which unions only represent about 13% of the overall workforce, and capital is globally mobile, the implementation of these four employment practices has compounded a situation of powerlessness—necessitating a new model for empowering workers.

A Model from the Past

The restaurant industry employs more than 10 million people, making it the largest private-sector employer in the nation. As such, it is a powerhouse of influence. The wages and standards it sets for workers have a heavy spillover effect on the rest of the US economy—especially the larger food industry. We've seen this spillover effect before in US history. At the turn of the last century, the auto industry was the largest private-sector employer in the nation, with over 6.5 million workers, and its low wages and harsh working conditions became the staple for the entire manufacturing sector. In 1933, according to a Federal Trade Commission report on the motor vehicle industry, the average wage for an autoworker was just over $1,000 per year, and by extension, the entire economy suffered. The low wages and dangerous working conditions dragged down the entire manufacturing sector and the economy overall. It was not until the United Auto Workers collectively bargained for better wages and conditions that an increase in spending, among other things, pulled the economy into recovery. As a matter of fact, the creation of the middle class of the 1950s and '60s was a direct result of the amplified voice employees had in the workplace.

ROC United's research has shown that, unlike the auto industry of days past, the restaurant industry is not a vehicle for the middle class . . . yet. However, we are at a crossroads. Restaurant owners can either buy into the old dog-eat-dog business model, or they can boldly embrace a "shared prosperity" model. The concept is simple: provide a healthy, locally grown, delicious meal and treat workers well; then they treat customers well and everyone wins.

Already, ROC United has won several victories, introducing a bill in Congress, developing groundbreaking research at the national level, and winning several more workplace justice campaigns.

A thriving business

Nationwide, and in each of the five regions studied by ROC United's recently released reports—New York City, Chicago, Metro Detroit, New Orleans, and the state of Maine—the restaurant industry is vibrant, resilient, and growing. In 2007, the restaurant industry contributed over $515 billion in revenue to the nation's Gross Domestic Product. Perhaps the industry's most important contribution to the nation's economy is the thousands of job opportunities and career options it provides. Despite the current economic recession, the restaurant industry continues to grow nationwide. In each locality, restaurant employment growth has outpaced that of the local region's economy overall. Since formal credentials are not a requirement for the majority of restaurant jobs, the industry provides employment opportunities for new immigrants whose prior experience outside the US may not be recognized by other employers; for workers that have no formal qualifications; and for young people just starting out.

In all five locations studied, we found two roads to profitability in the restaurant industry—the "high road" and the "low road." Restaurant employers who take the high road provide the best jobs in the industry, offering living wages, access to health benefits, and career advancement. Taking the low road to profitability, however, creates low-wage jobs with long hours, few benefits, and exposure to dangerous and often unlawful working conditions. Many restaurant employers in each of the five regions examined appear to be taking the low road, creating a predominantly low-wage industry in every region throughout the country, and making health and safety law violations commonplace.

While there are a few "good jobs" in the restaurant industry that offer living wages, the majority are "bad jobs" characterized by low wages, few benefits, and limited opportunities for upward mobility or increased income. According to the US Bureau of Labor Statistics (BLS), the national median hourly wage for food preparation and service workers in 2010 was only $9.54, including tips, which means that half of all restaurant workers nationwide actually earn less. In the same year, the federal poverty-line wage for a family of three was $8.86, meaning that more than half of all restaurant workers nationwide struggle in poverty.

At each location ROC studied, an overwhelming majority (over 90%) of restaurant workers surveyed reported that they do not have health insurance through their employers (see table 1). In addition, earnings in the restaurant industry have lagged behind those of the entire rest of the private sector. In terms of annual earnings, restaurant workers around the country made only $12,868 on average in 2008, compared to $45,371 for the total private sector, according to the BLS Quarterly Census of Employment and Wages (BLS 2009). A substantial number of workers in each local study reported overtime and minimum-wage violations, lack of health and safety training, and a failure to implement other health and safety measures in restaurant workplaces. In all five regions studied (New York, New Orleans, Chicago, Metro Detroit, and Maine), we found that workers of color hold most of the industry's "bad jobs," while white workers tend to disproportionately hold the few "good jobs." Last year, when I met two young African American workers stepping out into the humid night after a long shift in one of Bourbon Street's greasy pizza joints, they wasted no time in telling me they make

$4 an hour while the white women in the upscale oyster bar next door make $15 an hour with tips. "They'll never hire us over there. That's just the way it is," said Mickey, one of the two young men, as he lit up a smoke.

Workers also reported discriminatory hiring, promotion, and disciplinary practices. These challenges resulted in a three-dollar differential between white restaurant workers and workers of color in the five regions studied, with the median hourly wage of all white workers surveyed being $14.70, and that of workers of color being $11.50.

Table 1: Summary of Restaurant Workers' Experiences in Chicago, Metro Detroit, New Orleans, Maine, and New York City

PERCENTAGE OF WORKERS SURVEYED IN ALL AREA WHO ...	
Did not have health insurance provided through their employer	90.1%
Did not have paid vacation days	78.0%
Did not have paid sick days	89.6%
Worked while sick	66.7%
Suffered from overtime violations	38.3%
Reported that their experience of being passed over for a promotion based on race	30.9%
Reported having to do things under time pressure that might have harmed the health and safety of the consumer	24.4%
Reported that they or a family member had to go to the emergency room without being able to pay	26.5%
WAGE DIFFERENTIALS BY RACE	
Median Wage of white workers	$14.70
Median Wage of workers of color	$11.50

Source: ROC-United (2010)[1]

ROC research also sheds light on the hidden costs of low-wage jobs and low-road workplace practices to consumers, taxpayers, and the public at large. Violations of employment and health and safety laws place customers at risk and endanger the public. In

[1]Data has been weighted by position, industry segment, and size of local workforce.

each locality, we found that restaurant employers who violated labor laws were also more likely to violate health and safety standards in the workplace—such as failing to provide health and safety training, or forcing workers to engage in practices that harm the health and safety of customers. The pervasiveness of accidents, coupled with the fact that so few restaurant workers have health insurance, can lead to escalating, uncompensated care costs incurred by public hospitals. In all five localities, more than one-quarter (26.5%) of surveyed workers reported that they or a family member had visited the emergency room without being able to pay for their treatment.

Low wages and a lack of job security among restaurant workers leads to increased reliance on social-assistance programs, resulting in an indirect subsidy to employers engaging in low-road practices and fewer such public resources available to all in need. A key finding of our research was that whenever restaurant workers and high-road employers are hurt by low-road practices, so is the rest of society.

However, there is another path to profitability. It is possible to create good jobs while maintaining a successful business in the restaurant industry. Our interviews with employers revealed that as long as there is an enduring commitment to do so, it is possible to run a successful restaurant business while paying living wages, providing workplace benefits, ensuring adequate levels of staffing, providing necessary training, and creating career-advancement opportunities. In fact, in each locality, more than 10% of the workers we surveyed reported earning a living wage (in some locations, more than 20%), and similar numbers reported receiving benefits, thereby demonstrating both the existence of good jobs and the industry's potential to serve as a positive force for job creation. Workers who earn higher wages are also more likely to receive benefits, ongoing training, and promotion, and are less likely to be exposed to poor and illegal workplace practices. For example, workers earning a living wage, calculated according to locality, were also much more likely to have health insurance than workers earning less than the state minimum wage.

The ROC Model

The ROC model represents a new way to build power for workers through an entirely participatory process that surrounds the restaurant industry on all sides, using the three-pronged approach: research and policy, workplace justice, and promotion of the high road.

Research and policy

ROC engages in rigorous, statistically significant research of the restaurant industry that illustrates its workings in every market where a local ROC is located. The research utilizes a combination of surveys and interviews with a statistically representative sample of workers and employers that provide insight into the labor dynamics of their respective restaurant markets. But more importantly, the surveys allow us to meet workers and employers, develop relationships with them, and understand which issues run deep in the restaurant worker community.

Workers in local ROCs form policy committees that turn the issues workers care about into full-on policy campaigns aimed at improving the lives of all restaurant workers. Paid sick days, minimum wage, and other legislation would make conditions more tolerable for millions of restaurant workers.

Promotion of the high road

In every restaurant market there are ethically minded employers. They usually feel isolated and alone in the larger dog-eat-dog environment. The goal of the Restaurant Industry Roundtable is to bring these good folks together and allow them to share their best practices. The roundtable demonstrates that another model is not only possible but profitable. Additionally, roundtable restaurant owners will visit legislators side by side with workers, to advocate for legislation the local ROC policy committee has chosen to work on.

ROC Restaurant Industry Roundtables have developed a code of conduct that dictates what the basic standards and working conditions ought to be in the restaurant industry. In order to be a member of a roundtable, a restaurant must comply with this code and, in return, ROC promotes that restaurant in an ethical-eating consumer guide, and on its website and in print materials.

This model is not just for the restaurant industry but for the entire food chain. What happens in the restaurant industry affects not only workers, but everyone who eats. Food has become the epicenter of a new worldwide movement dedicated to creating a more just and sustainable food system. With our inherent ties to the food industry, restaurant workers are inextricably linked to it.

ROC United and the Food Movement

Three key questions have driven and synthesized this movement. The first question, "How does food affect the individual?" has led to a "green-food" revolution focused on the creation of good, healthy food that is free of pesticides and genetically modified organisms (GMOs). The energy and efforts put toward "good food" have led to a natural confluence with the second question, "How does food affect the environment?" Worries about GMOs, pesticides, and other such factors, including corporate seed domination, have generated a movement so powerful that the word *organic* is now a household term. In response to the third question, "What does food do to communities?" millions of poor people have cried out for food access, throughout both the developing world and marginalized communities in the developed world where food deserts are so prevalent. In some cases, it is a cry for access to "good food"—organics and other healthier choices—but in most cases, it is for access to *any* food. The issue is hunger. Some communities have acted on their own, setting up community gardens and other urban agricultural ventures.

Meanwhile, there is another question that has seldom been asked: How does food affect workers? Food does not magically appear in Whole Foods or on the plate of your favorite slow-food restaurant. There is a trajectory from farm to fork that involves workers at every step. From seed to harvest, shipping, warehousing, butchering, or processing; from preparing to serving at a restaurant, or stocking and selling at the local grocery store, workers do it all for us, the consumers. Workers are a fundamental part of the food chain. And workers and consumers are not in separate, isolated silos. Workers are consumers and vice versa. From production to consumption, the working class is the driving engine of the global food system.

Workplace justice

ROC holds low-road restaurants accountable via campaigns that use litigation and direct action to obtain improved conditions in the workplace. The campaigns focus on the most egregious labor and employment-law violators, which are abundant in the restaurant industry. Restaurants are particularly vulnerable to public pressure since they are such public institutions; something as simple as a party of five ROC sympathizers having dinner at a target restaurant can disrupt dinner service and cause such commotion that the business comes to a standstill. This makes outside-the-front-door activity, ranging from traditional picket lines to prayer vigils, extremely effective.

Our vision at ROC United is that all restaurant workers, regardless of where they work, will be able to eat healthy, sustainable food. The easiest way to improve access to good food in low-income communities is to raise the wages of those workers. Just as we did in the 1930s, we can once again create a new middle class—one that is engaged, conscious, and aware of the food they eat. The only way we can do this is by bringing workers wholly into the food justice movement and ensuring it remains class conscious, allowing for a critique that goes beyond our food to include the workers who make it. Food shaped my political understanding of the world and my family's place in this struggle. My hope is that food will create a world that does not require any more sacrifices from my family, or anyone else's, in order to simply live with bread and roses.

Works Cited

BLS (Bureau of Labor Statistics). May 2009. "National Occupational Employment and Wage Estimates United States. United States Department of Labor." Accessed March 13, 2011. http://www.bls.gov/oes/current/oes_nat.htm#35-0000.

ROC-United. 2010. "Behind the Kitchen Door; A Summary of Restaurant Industry Studies in New York, Chicago, Metro Detroit, New Orleans and Maine." Accessed June 6, 2011. http://www.rocunited.org/files/National_EXEC_edit0121.pdf

Chapter Twelve

WE EAT, WE DECIDE

By Xavier Montagut, President,
Xarxa de Consum Solidari, Barcelona, Catalunya

TODAY, FOOD PRODUCTION IS dominated by a handful of multinational companies that decide what we eat, how it is produced, and who produces it. The results are disastrous. Over a billion people suffer from hunger worldwide, thousands of farmers cannot produce food, and the paradox is that 80% of those going hungry belong to the rural population. The impoverishment of farmers is a global phenomenon. In the last 10 years Spain has lost an average of five farmers per day. The loss of fertile soil; water and soil pollution; the loss of biodiversity; and the impact on global climate are all characteristics of a dysfunctional, industrialized agrifood system. In terms of food quality, results are also terrible. Over 500 million people suffer from obesity while many more are under constant threat from mad cow disease, dioxide poisoning in chickens, and other ailments. This state of crisis urgently demands that citizens take control of our food system.

The prevailing agricultural model is annihilating the traditional family farming practices that have fed us for millennia while conserving the environment. That is why a worldwide farmers' movement is raising the banner of food sovereignty, rallying behind La Via Campesina, the largest coalition of farmers' organizations today.

However, La Via Campesina itself recognizes that farmers alone cannot overcome the modern food system. That is why they have proposed an alliance of farmers, consumers, environmentalists, NGOs, feminist organizations, and unions—an alliance among all popular social sectors. The call for unity has been issued at the Nyéléni 2007 Forum for Food Sovereignty (Declaration of Nyéléni 2007) and echoed by the Rural Platform (Plataforma Rural) in Spain.

An Alliance Takes Its First Steps

The Rural Platform has proposed a convergence and mobilization of food movements in order to transform the current agrifood system and establish a life-giving rural world. A veritable and resilient alliance requires a broad social base, one that is inclusive and participatory. It must be a balanced movement involving social movements and platforms, producers and manufacturers, consumers, NGOs, and environmental associations. It must represent everyone equally, avoiding discrimination based on gender, culture, territory, or age. Our goal is a movement capable of generating a broad mobilization toward food sovereignty through a common process of strengthening and articulating the diverse experiences that already exist in different places, and by framing them within a global strategy. It must be a space for political advocacy, dissemination, exchange, reflection, and shared debate.

I believe the alliance that is being built should have four characteristics. First, it should be a broad social movement that brings together all people, organizations, and entities that are working from the food sovereignty perspective. It requires a conscious effort to include all the different types of organizations (from local to international) to participate, independent of their size or resources or other characteristics (association, foundation, producer group, food co-op, NGO, etc.). Second, it must be a movement that maintains a balance between the different components of the alliance: producers, consumers, NGOs, ecological associations, etc., so as to reflect everyone's objectives, so that all feel represented in the alliance. This balance would be achieved by constantly working to reflect gender parity and to achieve a plurality of geographic representation. Third, it must be a movement that embodies the political accords defined in the Nyéléni documents, as it cannot otherwise hold together. At a time when the term *food sovereignty* has an ever-growing audience, it is important that these political agreements are honored. Fourth, it must result in social mobilization connecting the different forms of resistance (production, consumption, farmers' unions, political action, etc.) around a global strategy.

The People's Alliance for Food Sovereignty of the Iberian Peninsula

Throughout Spain, people are uniting behind the principle of food sovereignty (Alliance for Food Sovereignty 2011). Some new groups are going further than organizations traditionally viewed as the only ones with enough clout to have a substantial impact.

The Alliance for Food Sovereignty (Alianza por la Soberanía Alimentaria de los Pueblos; ASAP) has begun important work. A national meeting was held in February 2010, at which themes and priorities for the next year were established. A common agenda, action plan, and organizational structure were established. The approved plan clearly outlines the movement's objective to have a social and political impact that will form the backbone of the alliance.

The challenge is to achieve this impact while remaining equally inclusive of all representative groups. Accomplishing this requires long-term planning, proper timing, and methods that allow diverse organizations to participate—especially small groups and those with limited resources to act socially or politically. We must create an action calendar that allows decentralized work while demonstrating the alliance's strength. Ultimately, we need to mobilize in a way that produces concrete changes on the road to food sovereignty.

Healthy Food, a Consumer Demand

In Spain, it will be challenging to establish an ample social base around food sovereignty in a society where only 1.9% of the economically active population are farmers, and 10.75% of the population live in municipalities with fewer than 5,000 inhabitants. This means making food and agriculture a prime concern for a population that is largely separate from food production and agriculture.

We must look at what the interests are for citizens as consumers with regard to food sovereignty. With this in mind, it is in our favor that all the consumer trends increasingly reject the consequences of the current food model—and look for alternative ways to consume that are healthier, more respectful of the environment, and fairer to producers. We must educate the public about the dangers of a system that feeds us products containing high levels of dangerous substances without considering their cumulative effects on our health. Food security technocrats argue that food containing substances that have not been proven to be harmful to our health should not be prohibited. But why should we consume potentially harmful products when we can feed ourselves without them? We know the answer: it all depends on the agroindustry's determination to increase their profits. The good thing is that consumers' responses are informed and lead them in the opposite direction.

The Planet and Producers' Health

One of the modern food system's major impacts is a rapid loss of biodiversity. The Food and Agriculture Organization of the United Nations (FAO) estimates that 75% of crops' genetic diversity has been lost in the last century. We used to farm 7,000 to 10,000 species. Today, only 150 are cultivated. Only 19 crops and 8 animal species feed 95% of the modern world. In Spain, during the early 1970s, 380 varieties of melons were farmed. Currently, you can only find 10 to 12 species on the market. There are thousands of examples like this.

What criterion has determined which food varieties are used and which are abandoned? Nutritional characteristics? Taste? No. Rather, it is the ease with which they are industrialized and distributed. Long shelf life, controlled artificial ripening, the possibility of achieving large production with minimal costs, and external appearance are the prime characteristics of the corporate model's ideal fruit or vegetable.

The demand by consumers for better-quality food and respectful production methods is increasing the demand for organic products. The barometer of the perception and consumption of organic products, elaborated by the 2008 "Food and Ecological Agriculture Action Plan for Catalunya" (Generalitat of Catalunya 2011), shows an increase of 5% for those consuming organic products within the last year, and a 14% increase over the last three years. Among those interviewed, 92% believe it is positive to offer organic products in public eateries (hospitals, schools, etc.). In the cited survey, these consumers give special importance to healthy food that is in season, and is naturally- and locally-produced.

The dominant agrifood model has alarming social consequences as well. It is incapable of feeding over a billion people. Paradoxically, the majority of people suffering from hunger are farmers—many of them bankrupt. The industrial food model is the root cause of their ruin. Millions of farmers across the world have been forced to abandon their land and crops.

New Merchandise for New Demands? False Solutions

Large corporations have taken notice of increasing health concerns among consumers and wish to exploit it. They now offer merchandise that promises to replenish what we have lost through unbalanced diets and fast food—vitamins, fiber, calcium, and nutrients that help regulate cholesterol levels and have antioxidant properties. However, the marketing strategies behind food supplementation offer inaccurate information and half-truths.

For the more informed and demanding market, the agrifood industry has created a more sophisticated product. Large agrifood companies and supermarket chains have created organic product lines that follow new food-production guidelines (most importantly, a minimum of synthetic chemical products). However, these food items of improved quality are produced in the same industrial fashion, with high-energy consumption and degradation to the earth and water; and while they do take the well-being of animals into account, they do not consider that of the farmers who handle them. In other words, they have switched certain practices to ones that are less damaging to our bodies and the environment, but have not changed the essential method of production that destroys the planet, food quality, and farming communities.

In the same vein, "local products" have appeared next to natural products. The growing demand for local production on behalf of consumers stems from a desire to promote and protect local activities and local production networks. Local production is also important because it keeps money in the community, so that it has a dollar-multiplying effect and strengthens the local economy. Large industry, however, has reduced local production to an issue of miles, and fails to contemplate producers' characteristics or the stipulations of their contracts; they do not even contemplate the benefits that are generated for local economies. There is no limit to the extent to which large agrifood companies (especially large distributors) will go to exploit consumer demands, stripping them of their true meaning and simply adding a few characteristics to their products to increase their sales.

Products with "fair trade" labels are popping up all over supermarket aisles, even though it should be obvious that commerce is a relationship based on an exchange and should always be fair. These labels mask the true commercial mechanisms behind the products, referring only to the initial price paid to the grower. The current, unfair trade system is controlled by rules and relationships governed by large companies, especially large distribution chains, that benefit a select few at the expense of all other producers and consumers and the environment. To genuinely promote fair trade means to confront the system as a whole.

There are two lessons to be learned from these attempts by large corporations to appropriate the organic, the fair, the local, and the healthy. First, these schemes underscore the importance of the conscientious consumer and the needs of large corporations. Second, they emphasize the need to maintain a global vision of the modern food system in order to avoid being tricked into thinking a healthy, ecologically sustainable, and socially just diet can be achieved by new market niches that only benefit large corporations.

Significant Grotwth in Conscious Consumption

Multiple consumer organizations from various places have established a common cause. Thirty-one fair trade, responsible-consumption organizations have coordinated the "Space for Fair Trade" and declared, "The effort on behalf of fair trade organizations to import, distribute, and market products from popular organizations in the Global South is only part of the fight to change the unjust structures of international commerce..." They see their struggle as a fight "for food sovereignty in both the Global North and South. We consider it a strategic line that logically brings together our alternative as a group. We bet on agroecology against the industrialization of agriculture that channels all the benefits to megaindustries in the Global North" (Espacio Comercio Justo 2007).

Organic production lines have considered constructing alternative distribution networks that oppose currently dominant market mechanisms. We must evaluate these on the basis of sustainability. Weak networks revolve around products and focus on protecting their quality; strong ones focus on the process, addressing labor conditions, the role of rural communities, animal well-being, and conservation (soil, biodiversity, and water). In a study entitled "Identification and Topology for the Possibilities for Short-Circuit Commerce," Binimiles and Descombres (2010) synthesize the following criteria for defining strong distribution networks in Catalunya:

- The idea of proximity includes both the perspective of space (relocation) as well as the perspective of bringing consumer and production spheres closer together, with the purpose of resocializing these processes.
- Direct contact is the basis for relationships of trust and cooperation, and producers, manufacturers, consumers, and other actors have to have direct contact with one another.
- Information is understood to be the promotion of learning spaces and education about agrifoods, allowing decisions to be made in an autonomous and responsible manner, empowering consumers.
- Democratization through participation is the ability of people to directly participate in the governing and managing multiple components of the agrifood system.
- The fair redistribution of added value along the food chain is essential. Short-circuit alternatives seek economic behavior that allows us to capture the added value and get closer to reflecting the actual cost of production without losing the ability to include those sectors with less purchasing power.

Strong networks are growing in all aspects. The public services in Andalucía have developed a program to serve organic products in school cafeterias. In Catalunya, the Cafeteria Food Board exerts a strong influence on producers, environmentalists, consumers, and NGOs in the food-production process. Farmers' markets that were destroyed by large supermarkets or invaded by food carts with low-cost, imported, industrial agriculture products, are slowly reappearing with a strong organic component. A new wave of cooperatives and organic consumer groups has risen, producing a flux in distribution channels. In Catalunya in 2003, only 10 cooperatives were in practice; today, an estimated 130 exist. This means that there are about 4,000 families, and 12,000 people, participating. A similar growth process can be seen throughout the country.

The diversity of these alternative structures in food organization, distribution, consumption, and marketing is immense. Still, they share some common characteristics:

- They are projects based on trust between producers and consumers that maintain personal relationships stretching beyond commerce.
- They work with short distribution channels in terms of space (proximity), time (seasonal foods), and the distribution chain (eliminating unnecessary intermediaries).
- Pricing is established via a dialogue about necessities and possibilities for both consumer and producer, and not the ups and downs of speculative markets.
- These structures are stable and efficient in providing fresh garden products and fruits, the foundation of a Mediterranean diet.
- These mechanisms permit consumers and producers to participate in and control the organizations that have been created.

From Islands in an Archipelago to Knots in a Net

The emerging alternative to the modern food system can be characterized as an archipelago of isolated islands. It is critical that we overcome this fragmentation in the commercial and consumer realm if we wish to create a significant space for self-organization, consumer solidarity, and conscientious consumption that will empower the growing consumer sector to regain sovereignty over its food.

The recent proliferation of consumer groups is primarily thanks to the relative ease with which a small group can find a local farmer to fulfill basic food needs, but this basic approach increases the risk of paralyzing the ability of those groups to produce more substantial change and transformation. We find that once many consumer groups and cooperatives solve their basic problems, they either do not want or feel unable to challenge other aspects of the dominant model of consumption and distribution.

Certain issues are hampering the food movement's growth. One issue is the difficulty of analyzing costs in a clear and transparent manner. For example, volunteer work done by consumers and, in many cases by farmers, is not considered a cost. Until it is, many projects will remain constrained by small groups' limited free time.

The "volunteer" vision of self-management limits the reach of practical alternatives to the size of the current group. This is both the cause and effect of the fact that certain technical jobs are done in rudimentary fashion, which means that the little free time that exists is spent on doing things inefficiently. The true objective of self-management gets lost due to short-term vision and very inefficient use of work, resources, and socially achieved technologies. As a result, volunteer work that should be the source of replicable alternatives, and in turn a transformative vocation, becomes a limiting factor.

Another issue is the lack of an agroecological food culture. It is hard for farmers to establish a relationship with consumers that base their consumption on cosmetic appearance (the "eating with your eyes" approach used by the agroindustry in their marketing). It is now necessary to recover a food culture that should never have been lost.

From my point of view, the real challenge is for different types of alternative consumption organizations to fully complement each other. There are multiple ways of doing this; even within the same organization different functions can coexist. To create an intricate network of alternative-consumption cooperatives and associations requires imagination, flexibility, and a conscious effort to cooperate.

From Consumer Sovereignty to Food Sovereignty

While it is important to strengthen and promote alternative consumer commerce, we must not lose sight of its limitations. According to neoliberal theory, consumers steer the course of economics via the choices they make in what they buy and how. In a way, the purchase of an item is like a vote, expressing preferences among different options; all that matters is our "preferences" in choosing one product over another. In order to make a good decision, we simply have to have good information (the model demands "perfect information") and options (i.e., the absence of monopolies).

Assuming this neoliberal theory to be correct, when we buy from companies that act responsibly regarding the environment, women's rights, social rights, the poor, etc., we promote their values and contribute to social change. If our daily consumer behavior (buying and consuming are not synonymous) maintains the same moral values, we will eventually create a socially and environmentally responsible economy. In reality, however, economic systems are hard to change through consumer choice alone. For example, my transportation decisions are based on my spatial and seasonal restrictions: where I live, where I work, my schedule, my transportation options, etc. Most of these elements are outside my control. It is easier for changes in our economic system to influence our consumption habits than vice versa.

To improve our consumer conditions, we have to analyze what is called the *dictatorship of supply*. In many fields, the goods and services being offered are dominated by powerful capitalist groups that spend a huge amount of energy controlling our habits and shaping our environment around their interests (for example, destroying downtown areas by creating large leisure-shopping malls). Our consumer choices are often made based on conditions outside our control. There is also a distribution issue. As we all know, the market is a nondemocratic voting mechanism, in which a few have several votes, others have few votes, and still others have none. These inequalities outweigh the individual effect.

In the end, our social group often dictates our behavior, and we seldom see what is happening outside it. We have to realize that the ability to change the global economy through consumer behavior is very limited and we can't consider it the principal tool for change.

Citizens before Consumers: What to do?

More and more people are realizing we cannot continue with the suicidal dynamic of uncontrolled economic growth, and they would accept a more frugal lifestyle, knowing that they could live well, and maybe even better, with fewer material goods. However, individual habits, such as austerity, are not enough. We must acknowledge that changes in consumption stem from deep changes in the institutions and structures that regulate our economic life. Our most urgent task is to harness the good will of concerned citizens. We must extend citizenship into the consumption sphere (Sempere 2009). We need to make collective, democratic, political action a priority. We need to be citizens before we are consumers. Only then can we begin to create a different agriculture, a different consumption culture, and another world.

The present fragmentation of consumer groups and cooperatives clearly highlights problems that must be addressed. While many groups possess the will to transcend isolated work and join in collective political action, to transform that will into practical application is tremendously difficult. But the possibility has been demonstrated once before. In 2008–2009, the "We Are What We Plant" campaign (Som Lo que Sembrem)[1] gathered more than 100,000 signatures to pressure the Catalán Parliament to ban transgenic seeds in open fields. Organized consumers played an important role in the creation and diffusion of the campaign. Unfortunately, this momentum had no stability or continuity. The

[1] In 2009, Som Lo que Sembrem gathered 106,000 signatures to impede transgenic farming in Catalunya. The Catalán parliament, with a majority vote from both the right and the Socialist Party, negated discussing the proposal on July 2, 2009. More information can be found at http://www.somloquesembrem.org.

main challenge facing ASAP is to re-create this experience and create an action agenda capable of mobilizing consumer groups. The desire and will are there. The raw material, the wicker, is there. Let us hope there is enough wisdom to weave it into a basket, and that hope leads to action. That way, we can say that in Spain, food sovereignty has begun.

Works Cited

Alliance for Food Sovereignty, The. Accessed February 28, 2011. http://www.alianzasoberanialimentaria.org.

Binimelis, Rosa and Charles-André Descombes. 2010. "Comercialització en circuits curts: Identificació i tipologia. Escola Agrària de Manresa i Verloc. Amb la col·laboració de l'Associació l'Era." Barcelona. Accessed June 6, 2011. http://www.bagesformacio.cat/publicacions/Circuits%20Curts%20en%20Producci%C3%B3%20Agroaliment%C3%A0ria%20Ecol%C3%B2gica.pdf

Declaration of Nyéléni. 2007. World Forum on Food Sovereignty. Sélingué, Mali. Accessed February 28, 2011. http://nyeleni.org.

Espacio Comercio Justo. 2007. Accessed June 6, 2007. http://www.espaciocomerciojusto.org/index.php?option=com_content&view=article&id=2%3Amanifiesto-abriendo-espacio-por-un-comercio-justo&catid=25%3Aarticulos&Itemid=28&lang=e

Generalitat of Catalunya. Accessed April 1, 2011. http://www20.gencat.cat/docs/DAR/AL_Alimentacio/AL01_PAE/07_Pla_accio/Fitxers_estatics/Llibre_PdA_es.pdf.

Sempere, Joaquim. 2009. *Vivir Bien con Menos*. Colección Mas Madera. Barcelona: Icaria. Space for Fair Trade. Accessed March 1, 2011. http://www.espaciocomerciojusto.org/.

Chapter Thirteen

LOCAL FOODS ARE KEY TO ECONOMIC RECOVERY

By Ken Meter,
President, Crossroads Resource Center

BUILDING CLUSTERS of "local" foods businesses will be critical to economic recovery in the US. Without taking this step, we cannot put our national economy back on its feet. Now that the US has bailed out the banks, we have to transform our economy so it builds health, wealth, connection, and capacity in urban and rural communities. This will be especially true in farm regions and inner-city neighborhoods, which have been structurally weakened by economic exchanges. This chapter will show how some communities are nurturing local foods to launch an economic transformation of the US.

The bank bailout cost taxpayers hundreds of billions of dollars. This was probably an essential expense to avoid financial disruptions, yet the bailout left many of us with conflicting thoughts. It was certainly a way of rewarding those who had created a global fiscal crisis. Even more critically, the bailout left undone the very painstaking work of reversing our extractive economy. We will find that path through food. I say this for several reasons.

Finding a Softer Economic Path

First and foremost, food is a right: we have to ensure everyone eats well. This makes food very different from other products. If a manufacturer wants to sell a particular widget, and Dorothea can't afford it, most of us would agree that this is a transaction that will not occur. However, if Dorothea cannot afford food, we cannot simply say she is out of the market. We are required to understand why, and to respond by ensuring she has access to healthy food; this is especially true in a nation that prides itself on feeding the world.

In this way, we are literally forced to build softer relationships around food than we construct around other commerce. More than with any other product, trust is involved. When she can afford to eat, Dorothea has to trust that the people who raised, packaged, and delivered foods to her did their work with her health in mind. She cannot always tell from a package at the grocery store whether the item she buys will support her health; she has to get informed about each product, but in the end her decision to buy is based on trust. By the same token, farmers and meat packers and dairy bottlers that ship food to the market also have to trust that Dorothea (and millions like her) will choose to buy their products, because if she does not, those fresh products will spoil.

Second, everyone eats three times a day, if they are able to. Our relationship to food is more intimate than to other consumer items; we literally take food into our bodies. It is central to spirit, health, and well-being. If our eating is off-balance, this can affect everything we do. We have to get it right.

Third, the decisions we make around food are therefore central to our lives. We often make those decisions in consultation with others around us, and talk about our choices over, say, lunch. We are frequently processing information about nutrient content, price, freshness, and source. Our conversations and decisions about food create new meaning every hour—unlike those widgets.

Fourth, we spend a great deal on food. Food is the second-largest household expense, after housing. We spend a trillion dollars each year buying food—enough to pay for that bank bailout, or to halve the national debt. The daily decisions we make about food have a profound impact on the US economy.

Fifth, we have better public data sets about the agricultural economy than we have for manufacturing. This is because we understand that farmers play a central role in our economic and personal health. We ask them to report exceptional details about their businesses. Most firms would refuse to do this in the name of private enterprise. Yet farmers comply, and provide the Census of Agriculture considerable data about their profitability. This makes the food economy more transparent than many other sectors—and it also means we can learn about other economic sectors by examining food economies.

Sixth, our food economy is extremely efficient at taking wealth out of our communities. While farmers are not well rewarded for the new wealth they create, buyers, brokers, and processors have made fortunes by moving agricultural commodities. The concentrated power of the food-manufacturing industry means consumers pay higher prices than is necessary.

Seventh, a vibrant consumer movement has erupted in which consumers are expressing a strong desire to connect more directly with the source of their food—to know who produces their food, and to be assured it is safe, healthy food from a trusted source. Over the five-year period 2002–2007, food sales—from farmers directly to consumers—rose to $1.2 billion, a growth of 49%. This is a 10% increase each year. The number of farms selling directly also increased, from 116,733 to 136,817 over the same period, a gain of 17%. While this is only 0.4% of the commodities farmers sell, it is still a rising economic force while, at the same time, the fundamentals of the commodity economy have eroded, despite recent commodity price hikes.

Measured in terms of numbers of people and numbers of communities where people participate, this movement is already larger than the civil rights movement—and is still growing. Our collective hunger for better food has been acknowledged in the United States Department of Agriculture's (USDA) Know Your Farmer, Know Your Food campaign, launched in 2009.

Eighth, people are making drastically different choices about what they eat. Often, a cancer diagnosis persuades consumers to seek organic fruits and vegetables as part of their treatment. Other consumers choose fresh and organic produce, hoping to avoid illness. Farmers themselves often change their farming practices once chemical exposure threatens their families. If the goal of a food system is to build health, wealth, connection, and capacity in our communities, then our current system fails on all four counts. It fails precisely because it is an industrial commodity economy, more efficient at extracting wealth out of communities than ensuring people eat good, clean, fair food.

Failures of the Prevailing Food System

A few stark statistics tell a grim story. Health outcomes are dismal; food has become a leading cause of death, rivaling tobacco. A high-calorie diet, combined with a lack of exercise, accounts for one-fifth of the annual deaths in the US; 6 of the 15 leading causes of death are related to faulty diet and low physical activity.

The Centers for Disease Control and Prevention (CDC) report that two of every three US residents are overweight or obese (Flegal et al. 2010). The medical costs of treating obesity totaled $174 billion in 2009, the CDC adds (Finkelstein et al 2009). This staggering number is half the value of all food commodities currently sold by all farmers in the nation.

Poverty plays a key role in limiting nutrition. Though we may think we feed the world, half of all public school students in the US qualify for free or reduced-price school lunches because their families earn less than 185% of the income level at which the federal government officially considers a family in poverty. One out of every 10 households is uncertain at some point during the year where its next meal may come from, and is thus considered food insecure by the USDA. Farm income is frighteningly low, so rural communities suffer. Net cash income from farming crops and livestock in 2008—the best year since 1974, but only because grain speculators provoked a stiff rise in grain prices—was lower than in 1929 (after dollars are adjusted for inflation), following nine years of a rural depression. Once the speculative bubble burst, net cash farm income fell close to zero in 2009, a level similar to that of 1932, during the Great Depression.

Farmers collected an average of $19 billion each year in farm subsidies from 1999 to 2007. However, these subsidies do not mean rural communities are gaining wealth; quite the contrary. The overall impact of subsidies is to encourage farmers to keep farming, even though commodity prices do not cover their production costs. This is good for food buyers, who can buy at low prices. It is good for farm-input dealers, who can keep selling to farmers. It is good for lenders, since farmers continue to take on debt. However, one small measure of the imbalances created can be found in USDA Economic Research Service (ERS) data, which show that US farmers spent *$600 billion more* paying interest on farm debt than they received in subsidies from 1913 to 2007 (ERS farm income statement and balance sheet annual series 2007).

Truly, the food economy is extractive. Farmers face risks of climate, weather, and natural uncertainty that commodity brokers simply do not. Yet, when the ERS calculates the marketing bill for farm commodities, it finds that middlemen earn four dollars of revenue for each dollar earned by farmers (ERS food dollar series, annual series 2006). Moreover, the strength of the middlemen means that farmers are kept at quite a distance from consumers. Supply and demand cannot balance each other as we would like to believe—because the system disconnects them. Farmers respond to market signals they get from commodity buyers or federal programs, while consumers respond to marketing pitches from retailers. Social connections have been severed. Many of us live alone, sharing our meals with no one. Families find it difficult to eat meals together. Many children grow up with no adult showing them how to eat appropriate foods in moderate quantities, let alone how to cook. The age-old connections between culture and food have been weakened as people identify their food with corporate logos rather than with family, ethnic heritage, or sense of place.

Finally, we have less capacity than we once had in handling food safely. The most astounding measure of this breakdown of capacity is the loss of 3,000 lives each year in the US due to food poisoning—we essentially endure one 9/11 each year, created by our food supply. Furthermore, we spend $152 billion a year on the medical costs of treating food-borne illnesses. These are prominent signs that we do not know how to handle food with proper care. These statistics also suggest that state and national inspection systems, while precise in prescribing scientifically determined procedures, do not achieve the results we deserve. Several tests of food from family farms show less risk than food coming through an industrial processing system.

The Emerging Community-Based Foods Movement

Luckily, an alternative to the industrial food system is growing, and it is growing in every state of the nation. I am among those fortunate enough to touch base with this movement. Having worked to date with 63 regions in 27 states, and in one Canadian province, I can testify that this movement is unique in each place, because it cultivates the unique resources, skills, and qualities of each locale in which it emerges.

This movement is often called the local food movement, yet I have come to realize it is far more than that. For one thing, this movement does not suggest we build walls around our communities and insist that all foods eaten be produced within those walls. To insist on this would harm our own communities. Moreover, in most locales it is simply not practical.

Nor does this food movement count only the distance that food travels from farm to plate. Certainly, these "food miles" can be an interesting measure—yet so much more is involved. People want food that comes from a farm they know, yes, but they also want it to be raised in an environmentally sound way. They want the laborers that work in the fields to be paid adequate wages, and they want farmers to obtain a fair return for their labor. Overall, people want a sense of connection with their food supply.

In fact, careful eaters that are part of the local food movement often buy foods that are not local at all. Many would prefer food from a distant family farm to actual local food from an industrial operation. These people are interested in supporting community ownership and social connection, not simply in buying from whichever business happens to be closest. To me, this includes the possibility that I might buy bananas from a cooperative in Ecuador, or coffee from a shade-grown plantation in Guatemala, because these transactions have the quality of being local in the sense that I can know who the producers are, could contact them to be assured they are able to sell their products at fair prices, and to ask about their farming practices. This food is local in the sense that I can build a community-like connection with the farmer.

This movement has been more aptly called the "community-based food movement." I like this term, yet it can be difficult to use since few people recognize it. Let me define it: Community-based foods are produced in ways that build community connections. Community-based foods are not raised simply for profit—indeed, in many Native American communities food is considered a gift from the Creator, not a commodity to be sold. Those who uphold this belief are likely to insist on giving food away or on bartering. In more mainstream business contexts, community-based foods are supplied through "triple-bottom-line" business networks that strive to achieve financial goals—and more. These are firms that recognize their interdependence with their communities and try to build even greater connections based on mutual respect.

One community-minded business network

This has also been called relational commerce. This is still a difficult concept for some to grasp, so let me provide one example. A trusted colleague of mine is Mike Lorentz, co-owner (with his brother Rob) of Lorentz Meats in Cannon Falls, Minnesota. Mike and Rob purchased their parents' butcher shop, located in a town of 4,000 people about 45 minutes south of Minneapolis–

St. Paul, about 14 years ago. The business had been a respected main-street firm for three decades. Yet the brothers wanted to accomplish even more than their parents had. For one thing, they had two families to feed, not simply one. They also saw potential, years before many of us did, for delivering high-quality meats to consumers.

Mike and Rob took out substantial loans to expand their family operation. They opened up a new processing plant in the industrial park outside of Cannon Falls. They built up their custom-processing business, which was more profitable than commodity production. They designed and branded special ham, bacon, and sausages for national distribution. Over time, the Lorentzes also attracted high-quality processing work for Organic Prairie, the meat producers' cooperative. Now the Lorentzes handle most of the meat this firm trades in the Midwest. They attracted strong attention from investors, yet had difficulty finding any that wanted to help them expand in a way that left them free to grow the business patiently. Ultimately, they had to sell the original butcher business in order to expand their processing operation.

Now the Lorentzes employ 60 people. When I visit the shop, I see people working deliberately and diligently, seemingly without stress. The shop is always clean, and employees, many of whom are immigrants, exude a care for the business. When I interviewed Mike a couple of years back, he was willing to tell me a bit about the price spreads he maintains at the plant. For each pound of beef that moves through the building, Mike says he has to charge 35 cents. His mainstream competitors, by contrast, charge three cents per pound. Hearing this, some economists might suggest closing the factory. Yet Mike and Rob have found market niches that sustain the firm. With this cost structure, they are able to pay their employees well, produce a high-quality product, and expand the business.

But that was not enough to satisfy the Lorentzes. Community-minded, they also devote some of their business attention to helping foster the growth of other firms as well. For example, Mike once hired a consultant to help him figure out how to sell more grass-fed beef. Over time, the two of them realized the main obstacle to expanding this line was that the producers tended to be small and were scattered across the Midwest. It was costly to locate enough animals to build up the volume the plant needed to efficiently process grass-fed cattle. Ultimately, Mike helped the consultant spin off a new firm, which fills that exact function. By brokering cattle raised on small farms throughout the Midwest, this new firm was able to create sufficient volume to ensure that the meat plant could achieve better efficiency. This new firm, Thousand Hills Cattle Company, now brands its own grass-fed cuts, which are processed at Lorentz Meats and sold to restaurants, institutions, and grocers.

Nor was this the end of the story. Recently, a Latino-Anglo cooperative launched a free-range poultry operation in nearby Northfield, Minnesota. Their vision was to produce chickens intensively on small, quarter-acre plots. They see this as a viable strategy for helping immigrant families build savings and leading to farm ownership. Once they are able to produce enough chickens, the co-op hopes to build its own processing facility. Where did they turn for help distributing their chickens to Twin City markets? To Thousand Hills Cattle Company, which agreed to carry processed birds on their trucks.

This network of firms, then, has created both new opportunity and potentially new competition for Lorentz Meats. All these firms depend on each other in certain ways, and they combine efforts to make their region strong. Each firm builds community as well as a strong business. Networks like these are one example of what I mean by community-based foods.

When I interviewed Mike, I expected he might focus his attention on the close margins he had to cut to keep this delicate set of transactions in balance, and he certainly had some stories about this, yet he emphasized something completely different: the fact that to do business in this way, he needed to have trusting relationships with both his suppliers and his customers. If the trust were ever broken, the business could not thrive.

Strengthening the Local Multiplier

Clustering firms in this way within one locale accomplishes multiple benefits. For one, a community with, say, 10 food firms has 10 business owners that invest in the community rather than having most decisions made by one or two owners. By collaborating with each other, the firms help create stability for the entire cluster of firms. When they compete, they may induce each other to use resources efficiently. By having more employees in responsible positions, greater capacity is built among a more skilled workforce whose workers also have a wider range of work options than if they were merely employees. As these firms trade with each other, they also create local money flows that help recycle financial resources through the region. This serves to increase the economic multiplier—a measure of how many times a dollar is recycled through a given region once it is first earned.

In many discussions of economic development in American communities, the economic multiplier has become a key measure of a proposed new project's worth; powerful computer programs can grind out the calculations. Yet the concept of a multiplier is not widely understood. At core, it is a measurement of how frequently a business and its employees trade locally produced products. The more a firm sells local products, buys local inputs, and employs workers that also buy locally, the more the multiplier increases. Clearly, for the sake of increasing a multiplier, it is better to have several small firms trading with each other and exchanging with local consumers (as long as these local firms pay their workers well), than to have a few large firms managed from outside the community controlling economic transactions.

The more a community's businesses trade with each other, the longer a given dollar will linger. Some examples bear this out.

A Michigan study found a multiplier of 1.32 for produce raised on medium-sized family farms (Conner et al. 2008). For the state of Iowa, it was calculated that dollars spent at the state's farmers' markets cycled more, attaining a multiplier of 1.58 (Otto and Varner 2005). Another Iowa economist learned that a small restaurant that had committed itself to buying local foodstuffs generated a multiplier of 1.9, compared to a value of 1.53 for an average restaurant in the region (Enshayan 2008). An Oregon study found that each dollar spent buying food for school lunches cycled enough to create a multiplier of 1.87 (Ecotrust 2009). In one small-farm region of western Wisconsin, the multiplier for a small farm was calculated to be from 2.2 to 2.6 (Swain 2001).

For county commissioners or investors who find themselves waiting for the multiplier to rise before they invest, I offer this caution: if you want the multiplier to rise, you need to invest in making that happen by creating clusters of locally owned firms. This will in turn forge a more resilient local economy. The higher multipliers found in rural Iowa, Oregon, and Wisconsin are the result of over 40 years of community building, of creating strong social and economic networks. Often these locales have also had the wisdom to detach from the industrial system, relying upon themselves for essential services. We *construct* higher multipliers through public policy, infrastructure, and business decisions that favor the local.

Focus on Relationships

How do we get there? By focusing on building relationships of mutual respect, and learning to collaborate. Precisely because food systems are complex and rapidly changing, we need to have many eyes scanning the horizon from independent viewpoints. People from diverse cultures and with diverse experiences must work together to come up with strategies no single person could accurately pinpoint.

One of my best experiences in forging such collaboration has been in partnering with the Leopold Center for Sustainable

Agriculture, based at Iowa State University. The Leopold Center has worked diligently to reclaim the land-grant mission of the university, based upon the idea that the best expertise is forged when scholars and farmers work together to frame research and interpret its findings—not when experts define the choices that farmers can make.

Building connections in the inner city

In Flint, Michigan, patient work in a challenged urban setting is effectively building stronger social networks. One nucleus of activity has been Jacky and Dora King's karate school in inner-city Flint. After launching a successful business teaching these self-defense skills, Jacky began to articulate his sense that the neighborhood itself needed a self-defense strategy. Part of defending itself from economic upheaval was to produce more of its own food.

Jacky set up a hoop house, or hoop greenhouse, on an urban lot, and invited high school students to work the land with him. He assured them they would learn the skills of farming, and also gain work experience they could use in looking for jobs. These young urban farmers began to sell food at a recently revitalized farmers' market, as well as to local restaurants. They were supported by the Ruth Mott Foundation.

Gradually, an even wider vision began to emerge, as more and more people became involved and achieved more strength. These urban farmers are now convincing food-processing businesses to establish themselves near farmers' markets. This will foster more food production, and also give farmers a place to sell surplus crops. Jobs will be created in food processing. As these food businesses work with each other, the distance between producers and consumers will be reduced, while new capacities, such as leadership, will be built in the inner city. The city will gain greater financial stability as money cycles through different hands. They further envision a time when compost firms will build new soil for urban farms, making productive use of the urban food-waste stream and fall leaf accumulation.

This agricultural vision for the city has been boosted by a related civic investment launched years ago. A community land trust has been buying up vacant and foreclosed properties for years, as 16,000 autoworker jobs have been lost. Defunct houses are cleared off, and sod restored. This trust, which now owns 19% of all the land in the city, has made urban agriculture one of its priorities. These vast open areas have topsoil above and water mains below—irrigated land that could be farmed to feed nearby consumers that spend tens of millions of dollars on food each year.

Similarly, a collaboration of over 1,600 community gardeners has formed in nearby Detroit. The Garden Resource Collaborative emerged when residents met together to consider their vulnerability, as oil supplies peak, and their lack of access to healthy foods. Neighbors supported each other in developing gardens and irrigation systems. In so doing, they set up remarkable systems for cooperation. Gardeners readily grew surpluses beyond what their own families could use, and a cooperative distribution service was formed to sell these vegetables commercially; in its first year, 2010, $53,000 of produce was distributed by this co-op. Neighborhood resource centers now have been formed to help coordinate garden activity across the city.

Rural Food Initiatives Cluster Processing and Distribution with Food Production

Latino immigrants in Minnesota are working with their Anglo neighbors in rural areas to form a producer's co-op that will raise free-range chickens for Twin Cities markets. Regi Haslett-Marroquin, who spearheads the effort as a project within the Main Street Project, envisions dozens of scattered, quarter-acre chicken barns scattered across the countryside. Each would be built by hand, using simple wood frames covered in plastic sheeting to create a warm living space for the chickens. Using sprouted seeds, grasses, grains, and intensive management practices, Marroquin hopes to raise chickens in a way that will create ownership opportunities for immigrant families. Marroquin also envisions the co-op someday owning its own processing plant, to pack and ship birds to Twin City consumers.

In another example, a new hybrid co-op model is being launched through the inspiration of Rick Beckler, director of hospitality services for Sacred Heart Hospital in Eau Claire, Wisconsin. As food service director for a Franciscan hospital with a social mission, Beckler had been given a directive to purchase from local farms when possible. He began buying trout, buffalo, fruit, and vegetables from nearby producers. Yet few of the farms had returned to offer more products, and he wondered why. He finally confronted the growers at their annual winter value-added meeting, at which he said, "I'm spending two million dollars each year on food, and I'm tired of sending that money out of the region. I want to buy from you folks." He pointed out a couple of the farmers at the meeting, adding, "I've bought from you before. Why haven't you come back to sell me more?" The answer, it turned out, was that many of the farmers were inexperienced in marketing, and were too busy farming to focus on sales. Beckler realized that to fulfill his mission, he would have to help start a new intermediary that could convey food from these farms to the hospital, and to nearby schools and universities as well. Beckler is proud to be buying food from 21 local farms, but he does not have time to meet with each of them every week. He needed a brokerage that could convey food from all of these farms—and to keep in constant communication with all parties as markets fluctuated.

Margaret Bau, the co-op extension educator in Stevens Point, Wisconsin, suggested the group consider a hybrid model that had been widely adopted in Europe. In this model, the producers and buyers would join the same co-op board. They would invite their distributor and their trucker to join as well. Fair prices would be negotiated by all parties together, hoping to address a cluster of outcomes—that farmers and laborers were well paid, that environmentally sound practices were followed, and that buyers were charged affordable rates. Rather than set themselves up to have conflicting interests, they established reasons to collaborate. Now they are working together to build a stable and resilient distribution system. Sacred Heart Hospital has committed to spending 10% of its food budget through the co-op—and Beckler states that if the hospital attracts only two more patients a year that decide to seek health care there because they recognize the institution's commitment to local farmers, that will be enough extra business to cover the slightly higher food costs. Other institutional buyers are also being sought.

A couple of hours south, when a major national firm closed its factory suddenly in 2009, the town of Viroqua, Wisconsin, lost 85 of its best jobs overnight. The factory had been a profitable printing and packaging facility, but corporate owners decided they could save money by relocating the operation to a similar factory they already ran in another state. Yet this cost-saving gesture by the firm provoked a huge loss to the community. Sue Noble, director of economic development for Vernon County, called up the firm's CEO and asked, "What are you going to do for us, now that you have taken some of our best jobs out of town?" The executive was a bit befuddled by the request, but finally asked, "What would you like me to do?" Noble replied, "Sell us the building." He ultimately decided he would.

Now Vernon (County) Economic Development Association (VEDA) is the proud owner of a 100,000-square-foot building, certified clean enough for food processing, which will stand as the central institution of the region's food self-sufficiency effort. Two produce distributors have already decided to locate in the building, leasing equipment that will be held by VEDA. Hospitals and schools in the area have committed to purchasing food from the distribution center. Other food-processing and manufacturing businesses will locate in the former factory as conditions allow. Leaders researched the distribution options available to them and decided to start their own hybrid co-op, modeled after the Eau Claire co-op, mentioned above. The nearby $600-million co-op of co-ops, Organic Valley, has offered to be a partner. Setting up this storage and distribution system, leaders argue, will provide the infrastructure that will help make local food trade more efficient—and will simultaneously allow growers to have more reach into metropolitan markets in Chicago, Milwaukee, Madison, and the Twin Cities. The processing center was recently awarded a grant for $2 million from USDA to develop the building.

Framing Local Foods as Community Economic Development

A particularly inspiring story involves Vermont's Burlington Public Schools' effort to provide high-quality bread to their students in their daily sandwiches. Doug Davis and Robert (Bobby) Young, director and staff of the school nutrition program, approached an artisanal baker 50 miles away, asking if he could supply the school with enough bread to fulfill their needs. The baker asked them how many loaves of bread they needed each year, and how much they were willing to pay. When Doug told him, the baker at first demurred, saying he could not provide them bread at the price they were used to paying. However, as the baker went back to his baking, he thought over their proposal, and came up with a new idea. A couple of weeks later, he called Doug with a counteroffer: if the school could find a way to purchase 2,000 more loaves of bread each year, the baker could accept the price per loaf the school was proposing. At that volume, he could meet his costs. Doug and Bobby then reasoned that if they could get 100 of their school staff to buy one loaf of bread each week during the 20-week school year, they would indeed be able to order 2,000 more loaves of bread. Staff readily agreed. They could pay four dollars a loaf for artisanal bread delivered to their workplace—the school—on a weekly basis. Many of the staff shared the bread with their students in class; others took it home. These purchases made it possible for the district to buy the rest of the bread at a far lower wholesale price, giving the baker enough income to supply the school weekly.

The transaction actually earns money for the lunch program. With the proceeds, the district was able to buy grills for grilling sandwiches—and this, in turn, gave the baker a new market at the school, selling day-old bread that is better suited to grilling. Doug and Bobby consider this a community-supported bread operation, which shows that a connected community can find ways to be creative in problem solving, and in sharing risk.

Beyond Market Forces

It is critical to add that we can do this economically. Many of our policy decisions over the past several decades were based on a belief that natural market forces would guide us to better outcomes. The collapse of the financial system showed this belief to be false. Rather, a close investigation of the food system shows it to be more the outcome of public policy than of market forces. In the past we have offered tax incentives to those who expand their operations, and we have come to believe that larger firms are always more efficient—when in fact they may or may not be. We have built economic infrastructure (for example, financial institutions, highways, transportation routes, warehouses, and databases) that favor the large. We have allowed the large firms to prey off the small, even though more than 80% of all the businesses in the country hire less than 20 employees. We have come to believe that centralized livestock production and processing is more efficient, when in fact farm and ranch income has not been increased by these larger firms. In fact, farm income from raising livestock and related products (such as milk) fell, from $183 billion in 1969 to $150 billion in 2007 (after dollars are adjusted for inflation), while meat and milk consumption rose.

The plight of farmers is severe. Farmers' net cash income from producing crops and livestock was $21 billion lower in 2008 (the best farm year since 1974) than it was in 1969, despite the fact that farmers doubled productivity over that 39-year period. This means the risks farmers took on to become more efficient—assuming more physical risk by working faster, shouldering greater debt, living with more stress, and reducing costs—was a failed strategy from the farmer's standpoint. Indeed, perhaps we could use an agriculture that is less efficient, yet more rewarding to farmers.

If public policy played a key role in causing the dilemmas we now face, that is in many ways good news, because it means that public policy can help us build a better future. Certainly, we have to be aware of market realities—yet we often change market forces through policy. Public investment will be required to build a new infrastructure that creates local efficiencies in food storage, cooling, freezing, and transport. The more we build a food system based on green energy, the greater resilience we will have. In particular, the more our food system runs on locally produced green energy, the greater competitive advantage local networks will have relative to the fossil fuel–dependent industrial food sector. We will need public investment in local capacity building, and will have to create public incentives for local investment that cycles interest and dividend payments back into communities.

In this way, the word *local* resumes its essential truth and force. The measures of success for community-based foods include not only whether people get healthier, community connections are built, and community capacities are strengthened, but also whether local economies are strengthened. If locales are not getting stronger financially, then we have to refine our policy approaches, and learn new lessons about working and collaborating in communities, and in creating knowledge that fosters social cohesion.

This, then, is the face of economic recovery in the US. In community settings across the nation, far from the bailed-out banks that nevertheless refuse to make loans, and far from policy circles that believe communities are powerless in this age, creative grassroots efforts like those outlined here are using food to grow new options for America, built on social networks and business clusters.

Works Cited

Conner, David S. et al. 2008. "The Food System as an Economic Driver: Strategies and Applications for Michigan." *Journal of Hunger & Environmental Nutrition* 3 (4): 371–83.

Ecotrust. 2009. "Farm to School Investment Yields a Healthy Return into State Coffers." Accessed August 23, 2010. http://www.ecotrust.org/press/f2s_investment_20090318.html.

See Enshayan, Kamyar. (2008). Community Economic Impact Assessment for a Multi-County Local Food System in Northeast Iowa. Leopold Center for Sustainable Agriculture Final Report M08-05. Summary viewed July 7, 2011 at http://www.leopold.iastate.edu/research/marketing_files/food/food.htm.

Swenson, Dave. 2007. "Economic Impact Summaries for Local Food Production." Iowa State University: Leopold Center for Sustainable Agriculture, and University of Northern Iowa Center for Energy and Environmental Education, unpublished, March.

Finkelstein, EA, Trogdon, JG, Cohen, JW, and Dietz, W. Annual medical spending attributable to obesity: Payer and service-specific estimates. *Health Affairs* 2009; 28(5): w822-w831.

Flegal, Katherine M. Ph.D., and Margaret D. Carroll, MSPH, Cynthia L. Ogden, PhD. Lester R. Curtin, PhD. 2010. Prevalence and Trends in Obesity Among US Adults, 1999-2008. Journal of the American Medical Association. (303) 3. Accessed June 6, 2011. http://jama.ama-assn.org/content/303/3/235.full.pdf+html

Meter, Ken (2006). "Evaluating Farm and Food Systems in the U.S.," in Williams, Bob and Iraj Imam, eds. (2006). *Systems Concepts in Evaluation: An Expert Anthology.* American Evaluation Association monograph published by EdgePress of Inverness, 146.

Otto, Daniel, and Theresa Varner. 2005. "Consumers, Vendors, and the Economic Importance of Iowa Farmers' Markets: An Economic Impact Survey Analysis." Iowa State University: Leopold Center for Sustainable Agriculture and its Regional Food Systems Working Group, Part of the Value Chain Partnerships for a Sustainable Agriculture, Report RWG03-04, March. Accessed March 24, 2011. http://www.iowaagriculture.gov/Horticulture_and_FarmersMarkets/pdfs/FarmMarketReportMarch2005.pdf.

Swain, Larry. Interview with economics professor Larry Swain, community development specialist for the University of Wisconsin Extension Service and director of the Survey Research Center at UW-River Falls, [now retired] February 12, 2001

See Swain, Larry (1999). "A Study of the Economic Contribution of Small Farms to Communities – Completed 1996 to 1999." Unpublished manuscript; and Swain, L. B., & Kabes, D. (1998). "1996 Community Supported Agriculture Report." Unpublished manuscript.

Meter, Ken and Rosales, Jon (2001). *Finding Food in Farm Country,* pages 19-20. Community Design Center, University of Minnesota Experiment in Rural Cooperation, and Crossroads Resource Center. Available at http://www,crcworks.org/ff.pdf

United States Department of Agriculture Economic Research Service, Farm Income Statement and Balance Sheet data. http://ers.usda.gov/Data/FarmIncome/finfidmu.htm. Calculated by the author using constant 2007 dollars. An earlier compilation was reported in the literature in Meter, 2006.

United States Department of Agriculture Economic Research Service. Food dollar series. (Annual series). http://www.ers.usda.gov/Data/FoodDollar/index.htm. This series replaces the "Marketing Bill" series in effect when this chapter was written. Former address is http://www.ers.usda.gov/data/farmtoconsumer/marketingbill.htm.

PART THREE:

DEVELOPMENT,
CLIMATE
AND
RIGHTS

Chapter Fourteen

THE TRANSFORMATIVE POTENTIAL OF AGROECOLOGY

By Olivier De Schutter,
UN Special Rapporteur on the Right to Food

AGRICULTURE *IS* AT A CROSSROADS. For almost 30 years, since the early 1980s, neither the private sector nor governments have shown interest in agricultural investment. This is now changing. Over the last few years, agrifood companies have increased direct investment as a means to lower costs and ensure the long-term viability of supplies. The global food-price crisis of 2007–2008 also pushed governments into action. In July 2009, the G8 Summit in L'Aquila produced a food security initiative, promising to mobilize $20 billion to strengthen global food production and security; and the Global Agriculture and Food Security Program (GAFSP) was established as a multilateral financing mechanism to help implement these pledges. Other initiatives at global and regional levels are underway, such as the Comprehensive Africa Agriculture Development Programme (CAADP), created by New Partnership for Africa's Development (NEPAD). Governments are paying greater attention to agriculture than in the past. The "urban bias" is still very present (Lipton 1977), as most governmental elites still depend on political support from urban populations, but the prejudice against agriculture is diminishing.

However, recent efforts to combat hunger and malnutrition will fail if they do not produce higher incomes and improved livelihoods for the poorest—particularly small-scale farmers in developing countries. And short-term gains will be offset by long-term losses if ecosystems are further degraded, threatening future ability to maintain current levels of production. The question therefore is not simply *how much*, but also *how*. Pouring money into agriculture will not be sufficient; we need to take steps that facilitate a transition toward a low-carbon, resource-preserving type of agriculture that benefits the poorest farmers.

The Right to Food as a North/South Convergence Strategy: The US Experience

By Molly D. Anderson (College of the Atlantic)

Molly Anderson begins her discussion on the right to food as a North/South convergence strategy by launching into the fact that the right to food is recognized by almost all nations, yet is violated with alarming frequency. Anderson explains that in order to eliminate food insecurity in the US, the right to food has to be better understood. Of almost equal importance, this understanding could bridge a gap between US civil society and the international movement for food sovereignty, which includes the right to food as its first principle.

Both the right to food and food sovereignty are marginal and marginalized by both the US government and US civil society, despite some increasing interest in food sovereignty. Even among US organizations that recognize or accept the right to food, actions to implement it are not given priority. Food security is stripped of the important political questions of who has power; how and by whom food is produced; and who is served first. The failure to acknowledge that these questions are vitally important is a huge barrier in the way of universal food security—and a powerful impetus to the international movement for food sovereignty.

Anderson emphasizes that a rights-based approach to development is integral to a North/South convergence strategy. A rights-based approach entails identifying root causes of poverty, empowering rights-holders to claim their rights and enabling duty-bearers (usually governmental) to meet their obligations. Organizations with the strongest connections to peasant organizations in other countries, such as La Via Campesina, or that use farmer-to-farmer connections in their work, are the strongest advocates of food sovereignty. North/South interests will remain in conflict until the Global North can more effectively engage with the Global South to figure out how to meet basic human rights for all.

Full Article at: http://www.foodmovementsunite.com/addenda/anderson.html

Agroecology can play a central role in achieving this goal. We can significantly improve agricultural productivity where it has been lagging behind, and thus raise production in poor, food-deficit countries, while preserving ecosystems and improving the livelihoods of smallholder farmers. This would slow the trend toward urbanization in volatile countries, which is putting stress on their public services. It would contribute to rural development and ensure the succeeding generation can meet its own needs. And it would contribute to the growth of other economic sectors, as higher incomes in rural areas would stimulate demand for nonagricultural products.

A Diagnosis

The global food-price crisis has primarily focused attention on increasing overall production. The crisis has largely been attributed to a mismatch between supply and demand, a gap between slower productivity growth and increasing needs. A widely cited estimate states that, taking into account demographic growth, as well as changes in the composition of diets and consumption levels associated with increased urbanization and higher household incomes, there should be a 70% increase in overall agricultural production by 2050 (Burney et al. 2010).

We should treat this estimate with caution. First, it takes the current demand curves as given. Meat consumption is predicted to increase from 82.28 lbs/person/year in 2000 to over 114.4 lbs/person/year by 2050, so that by mid-century, 50% of total cereal production may have to go to increasing meat production (FAO 2006). The United Nations Environmental Programme (UNEP) estimates that, even accounting for the energy value of the meat produced, the loss of calories incurred from feeding cereals to animals instead of people equals the annual caloric need of over 3.5 billion people (UNEP 2009). In addition, as a result of agrofuel promotion, the diversion of crops from meeting food needs to meeting energy needs tightens the pressure on agricultural supplies.

Second, waste in the food system is considerable. For instance, the total amount of fish lost through discards, postharvest, loss and spoilage may be around 40% of landings (Akande and Diei-Ouadi 2010). Food losses in the field due to pests and pathogens may be as high as 20%–40% of potential harvests in developing countries, and average postharvest losses resulting from poor storage and conservation amount to at least 12%–50% (UNEP 2009).

Third, even though food availability may have to increase, we must remember the main cause of hunger today is not low stocks or global supplies unable to meet demand, but poverty. Increasing the incomes of the poorest is essential to ending hunger. We need to invest in agriculture, not only in order to match growing needs, but also to reduce rural poverty. Because poverty remains so heavily concentrated in rural areas, GDP growth originating in agriculture has been shown to be at least twice as effective in reducing poverty as GDP growth in other sectors (World Bank 2008). Only by supporting small producers can we help break the vicious cycle that leads from rural poverty to the expansion of urban slums, in which poverty breeds more poverty.

Fourth and finally, agriculture must not compromise its ability to satisfy future needs. The loss of biodiversity, unsustainable use of water, and pollution of soils and water compromise the continuing ability of natural resources to support agriculture. Climate change, which translates into more frequent and extreme weather events such as droughts and floods and less predictable rainfall, is already impairing the ability of certain regions and communities to feed themselves, and destabilizes markets. The change in average temperatures is threatening the ability of entire regions, particularly those living from rain-fed agriculture, to maintain actual levels of agricultural production (Stern 2007). Less fresh water will be available for agricultural production, and the rise in sea level is already causing the salinization of water in certain coastal areas, making water sources improper for irrigation purposes. By 2080, 600 million additional people could be at risk of hunger, as a direct result of climate change (UNDP 2007).

The current development path of industrial agriculture is worsening this situation. Agriculture accounts for at least 13%–15% of global, man-made greenhouse gas (GHG) emissions (Kasterine and Vanzetti 2010). In addition, the GHG intensity of agriculture increases faster than its productivity: while agricultural emissions of methane and nitrous oxide grew by 17% between 1990 and 2005, cereal yields increased by only 6% over the same period (Hoffman 2010). In other words, industrial agriculture is becoming *more* carbon intensive. Without a substantial change in policy, the GHG emissions from agriculture could rise by 40% by 2030 (Smith et al. 2007).

Agroecology is gaining recognition as a way to address these challenges among an increasingly wide range of scientific experts (McIntyre et al. 2009) and international agencies such as the United Nations' Food and Agriculture Organization (FAO), Bioversity International, and UNEP. It is also gaining ground in countries as diverse as the United States, Brazil, Germany, and France.

Agroecology: A Solution to the Crisis of Food Systems?

Agroecology has been defined as the "application of ecological science to the study, design and management of sustainable agroecosystems" (Altieri 1995). It improves agricultural systems by mimicking or augmenting natural processes, thus enhancing beneficial biological interactions and synergies among the components of agrobiodiversity. Common principles of agroecology include recycling nutrients and energy on-farm, rather than relying on external inputs; integrating crops and livestock; diversifying species and genetic resources in agroecolosystems over time and space, from the field to landscape levels; and focusing on interactions and productivity across the agricultural system, rather than focusing on individual species. Agroecology is highly knowledge intensive, based on techniques that are not delivered top-down but developed on the basis of farmers' knowledge and experimentation. The diversity of species involved in agroecological practices (including animals) requires a diversification of farm tasks.

The techniques involved in agroecology have been developed and successfully tested in many regions (Pretty 2008). Integrated nutrient management reconciles the need to fix nitrogen in farm systems with importing inorganic and organic sources of nutrients and the reduction of nutrient losses through erosion control. Agroforestry incorporates multifunctional trees into agricultural systems; in Tanzania, 350,000 hectares of land have been rehabilitated through agroforestry in the western provinces of Shinyanga and Tabora (Pye-Smith 2010). Similar large-scale projects are underway in other countries, including Malawi, Mozambique, and Zambia (Garrity et al. 2010). Water harvesting in dryland areas allows for the cultivation of formerly abandoned and degraded lands, and improves the water productivity of crops. In West Africa, stone barriers built alongside fields decelerate and stop runoff water during the rainy season, allowing an improvement of soil moisture, the replenishment of water tables, and reductions in soil erosion. Additionally, water retention capacity is multiplied five to tenfold, the biomass production multiplies by 10–15 times, and livestock can feed on the grass that grows along the stone barriers (Diop 2001). The integration of livestock into farming systems, such as dairy cattle, pigs, and poultry, and using zero-grazing cut-and-carry systems, provides a source of protein for the family as well as a means of fertilizing soils; so does the incorporation of fish, shrimps, and other aquatic resources into farm systems such as irrigated rice fields and fish ponds. These approaches involve the maintenance or introduction of agricultural biodiversity (the diversity of crops, livestock, agroforestry, fish, pollinators, insects, soil biota, and other components that occur in and around production systems) to achieve the desired results in production and sustainability.

Sometimes apparently minor innovations can provide high returns. In Kenya, researchers and farmers developed the "push-pull" strategy to control parasitic weeds and insects that damage crops. The strategy has been promoted in particular by Hans Herren, another contributor to this volume. It consists in "pushing" away pests from corn by interplanting it with insect-repellent crops like *Desmodium*, while "pulling" them toward small plots of Napier grass, a plant that excretes a sticky gum that both attracts the pest and traps it. The system not only controls the pests but has other benefits as well, because *Desmodium* can be used as fodder for livestock. The push-pull strategy doubles maize yields and milk production while improving soils at the same time. The system has already spread to more than 10,000 households in East Africa by means of town meetings, national radio broadcasts, and farmer field schools (Khan et al. 2011). In Japan, farmers found that ducks and fish were as effective as pesticides in rice paddies for controlling insects, while providing additional protein for families. The ducks eat weeds, weed seeds, insects, and other pests, thus reducing weeding labor, otherwise done by hand by women. Duck droppings provide plant nutrients, and duck swimming activity increases rice growth, leading to stockier stems. The system has been adopted in many rice-growing areas in Bangladesh, China, India, and the Philippines. In Bangladesh, the International Rice Research Institute reports 20% higher crops yields and net-income increases on a cash-cost basis of 80% (Mele et al. 2005).

Such resource-conserving, low-external-input techniques have a huge, yet still largely untapped potential to address the combined challenges of production, of combating rural poverty and contributing to rural development, and of preserving ecosystems and mitigating climate change.

Agroecology as a response to the question of supply

Agroecological techniques have a proven potential to significantly improve yields. In what may be the most systematic study of their potential to date, Jules Pretty et al. (2006) compared the impacts of 286 recent sustainable agriculture projects in 57 poor countries, covering 37 million hectares (3% of the cultivated area in developing countries). They found that such interventions increased productivity on 12.6 million farms, with an average crop increase of 79%, while improving the supply of critical environmental services. Disaggregated data from this research show that average food production per household rose by 1.7 tons per year (up by 73%) for 4.42 million small farmers growing cereals and roots on 3.6 million ha, and by 17 metric tons per year (up 150%) for 146,000 farmers on 542,000 hectares cultivating roots (potato, sweet potato, cassava). After the United Nations Conference on Trade and Development (UNCTAD) and UNEP reanalyzed the database to produce a summary of the impacts in Africa, it was found that the average crop-yield increase was even higher for these projects than the global average (79%): a 116% increase for all African projects and a 128% increase for the projects in East Africa (UNCTAD and UNEP 2008).

The most recent large-scale study points toward the same conclusions. Research sponsored by the UK's Government Office for Science (2011) reviewed 40 projects in 20 African countries where sustainable intensification was developed during the first decade of this century. The projects included crop improvements, particularly through participatory plant breeding on hitherto neglected orphan crops, and integrated pest management, soil conservation, and agroforestry. By early 2010, these projects had documented benefits for 10.39 million farmers and their families, and improvements on approximately 12.75 million hectares. Crop yields more than doubled on average (increasing 2.13-fold) over a period of three to 10 years, resulting in an increase in aggregate food production of 5.79 million tonnes per year, equivalent to 1,225.4 lbs per farming household. It should be noted, however, that not all of these projects fully comply with the principles of agroecology.

Agroecology's ability to increase the incomes of small-scale farmers

One advantage of agroecology is its reliance on locally-produced inputs. Many African soils are nutrient poor and heavily degraded, needing replenishment. But supplying nutrients to the soil can be done not only by applying mineral fertilizers, but also by applying livestock manure or growing green manures. Farmers can also establish what has been called "a fertilizer factory in the fields" by planting trees that take nitrogen out of the air and "fix" it in their leaves, which are subsequently incorporated into the soil. A tree such as *Faidherbia albida*, a nitrogen-fixing acacia species indigenous to Africa and widespread throughout the continent, performs such a function (ICRAF 2009).

The use of nitrogen-fixing trees avoids dependence on synthetic fertilizers, the price of which has been increasingly high and volatile over the past few years, exceeding the price of food commodities even when the latter reached a peak in July 2008. This means that whatever financial assets a household has can be used on other essentials, such as education or medicine.

Agroecology diminishes the dependence of farmers on external inputs and thus on subsidies, the local retailer of fertilizers or pesticides, and the local moneylender. Diversified farming systems produce their own fertilizers and their own pest control, thus diminishing the need for pesticides (Altieri and Nicholls 2004). The availability of adapted seeds, planting materials, and livestock breeds also presents multiple advantages for the farmer, while providing the required diversity of materials in major crops such as maize, rice, millet, sorghum, potato, and cassava. This is particularly beneficial to small-scale farmers (especially women) with low or no access to credit, no capital, or whom fertilizer distribution systems often do not reach, particularly since the private sector is unlikely to invest in remote areas where communication routes are poor and few economies of scale can be achieved.

A study on agroforestry in Zambia, involving intercropping or rotation between various trees and maize, showed that the net benefit of agroforestry practices is 44%–58% superior to nonfertilized, continuous maize production practice. And while subsidized, fertilized maize was the most financially profitable of all the soil fertility management practices, subtracting government's 50% subsidy on fertilizer sharply reduces the difference in profitability between fertilized maize and agroforestry, from 61% to 13%. Even more importantly, agroforestry practices yielded higher returns per unit of investment cost than continuous maize fields with or without fertilizer. Each unit of money invested in agroforestry practices yielded returns ranging between 2.77 and 3.13 (i.e., a gain of between 1.77 and 2.13 per unit of money invested) in contrast with 2.65 obtained through subsidized fertilized maize, and 1.77 through nonsubsidized fertilized maize. The return to labor per person per day was consistently higher for agroforestry practices than for continuous maize practice. The study noted that "in rural areas where road infrastructure is poor and transport costs of fertilizer are high, agroforestry practices are most likely to outperform fertilized maize in both absolute and relative profitability terms" (Ajayi and Akinnifesi 2007).

The contribution of agroecology to rural development and other sectors of the economy

Agroecology contributes to rural development because it is fairly labor-intensive and most effectively practiced on relatively small plots of land. The launching period is particularly labor-intensive because of the inherent complexity in managing different plants and animals, and of recycling the waste produced. Still, research shows that the labor-intensity of agroecology in the long term may have been exaggerated (Ajayi and Akinnifesi 2007). In addition, while governments have generally prioritized labor-saving policies, increasing employment in developing countries' rural areas, where underemployment is currently massive and demographic growth remains high, may constitute an advantage rather than a liability, and may decelerate rural–urban migration.

Although they can create jobs, agroecological approaches are fully compatible with a gradual mechanization of farming. In fact, the need to produce equipment for conservation-agricultural techniques such as no-till and direct seeding could create jobs in the manufacturing sector. This is particularly true in Africa, which still imports most of its equipment, but increasingly manufactures simple equipment such as jab planters, animal-drawn planters, and knife rollers. Employment could also result from the expansion of agroforestry. In southern Africa, farmers produce trees as a business, supported by a financing facility established by the World Agroforestry Centre (ICRAF). During its first year, the Malawi Agroforestry Food Security Program distributed tree seeds, setting up 17 tree nurseries that raised 2,180,000 seedlings and establishing 345 farmer groups (Pye-Smith 2008).

Growth in agriculture can be especially beneficial to other economic sectors if it is broad based, increasing farming household incomes rather than further enriching large landowners that rely on large-scale, heavily mechanized plantations. One argument states that growth in agriculture benefits other sectors because it raises input demand and leads to growth in agroprocessing activities, respectively upstream and downstream from the production process on the farm. More importantly, however, increased incomes in rural areas will raise demand for locally traded goods or services. This is particularly likely where agricultural growth is widely spread across large segments of a very poor population (Christiaensen 2011).

Agroecology's contribution to improving nutrition

Green revolution approaches in the past have focused primarily on boosting cereal crops (rice, wheat, and maize) in order to avoid famines. However, these crops are mainly a source of carbohydrates. They contain relatively little protein and few of the other nutrients essential for adequate diets. The shift from diversified cropping systems to simplified, cereal-based systems thus contributed to micronutrient malnutrition in many developing countries (Demment et al. 2003); of the over 80,000 plant species available to humans, only three (maize, wheat, and rice) currently supply the bulk of our protein and energy needs (Frison et al. 2006). Nutritionists increasingly insist on the need for more diverse agroecosystems to ensure a more diversified nutrient output.

The diversity of species on agroecological farms, as well as in urban or periurban agriculture, is an important asset in this regard. For example, it has been estimated that indigenous fruits contribute on average about 42% of the natural food basket rural households rely on in southern Africa (Campbell et al. 1997). Not only is this an important source of vitamins and other micronutrients; it may also be critical for sustenance during lean seasons. Nutritional diversity is of particular importance to children and women.

Agroecology and climate change

Agroecology supports the health of our ecosystems. It provides a habitat for wild plants, supporting genetic diversity and pollination, and water supply and regulation. It also improves resilience to climate change. Climate change means more extreme weather-related events. The use of agroecological techniques can significantly cushion the negative impacts of such events; resilience is strengthened through agricultural biodiversity (The Christensen Fund and Bioversity International 2010). As Eric Holt-Giménez (2002) has shown, following Hurricane Mitch in 1998, farming plots cropped with simple agroecological methods (including rock bunds or dikes, green manure, crop rotation, and the incorporation of stubble, ditches, terraces, barriers, mulch, legumes, trees, plowing parallel to the slope, no-burn, live fences, and zero-tillage) had on average 40% more topsoil, higher field moisture, less erosion, and lower economic losses than control plots on conventional farms. On average, agroecological plots lost 18% less arable land to landslides than conventional plots, and had a 49% lower incidence of landslides, and 69% less gully erosion.

More frequent and severe droughts and floods can be expected in the future, and agroecological modes of farming are better equipped to handle them. The agroforestry program developed in Malawi protected farmers from crop failure after droughts, thanks to the improved soil filtration it allowed (Akinnifesi et al. 2010). Indeed, on-farm experiments in Ethiopia, India, and the Netherlands have demonstrated that órganic farm soils' physical properties improved drought resistance in crops (Eyhord et al. 2007). In Brazil, a sixfold difference was measured between infiltration rates under low-tillage agriculture and industrial tillage. The former allows rainfall to better recharge groundwater and reduces the risk of flooding (Landers 2007). The soil's infiltration capacity is also maintained by the use of mulch cover, which protects the soil surface from temperature changes and minimizes soil evaporation (Kassam et al. 2009). In addition, agroecology's diversity of species and farm activities mitigates risks posed by extreme weather events, as well as those posed by the invasion of new pests, weeds, and diseases sure to result from global warming. Several agroecological approaches, such as cultivar mixtures, increase crop heterogeneity and genetic diversity in cultivated fields. This improved crop resistance to biotic and abiotic stresses in the Yunnan Province in China; after disease-susceptible rice varieties were planted in mixtures with resistant varieties, yields improved by 89% and blast (a major disease in rice) was 94% less severe than when they were grown in monoculture, leading farmers to abandon the use of fungicidal sprays (Zhu et al. 2000).

Agroecology also puts agriculture on a path to sustainability, by delinking food production from our reliance on fossil energy (oil and gas). And it contributes to mitigating climate change, both by increasing carbon sinks in soil organic matter and aboveground biomass, and by reducing greenhouse gases (GHGs) through direct and indirect energy use. The Intergovernmental Panel on Climate Change (IPCC) (2007) has estimated the global technical mitigation potential for agriculture to be 5.5–6 Gt of CO_2-equivalent per year by 2030. Of this, 89% can come from carbon sequestration in soils by storing carbon as soil organic matter (humus); 9% from methane reduction in rice production and livestock/manure management; and 2% from nitrous oxide reduction through better cropland management (Hoffman 2009).

Scaling Up Agroecology

The discussion above points to the need for an urgent reorientation of agricultural development toward systems that use fewer external inputs linked to fossil energies, and that use plants, trees, and animals in combination, mimicking nature instead of industrial processes at the field level. However, in moving toward more sustainable farming systems, time is the greatest limiting factor; success largely depends on our ability to learn faster from recent innovations and to disseminate what works more widely.

Governments have a key role to play in this regard. A shift toward sustainable agriculture entails transition costs, since it requires that farmers learn new techniques. A successful transition largely depends on the farmers themselves taking the lead. Governments should encourage learning from farmer to farmer, in farmer field schools, or through farmers' movements such as the *Campesino-a-Campesino* movement in Central America and Cuba (Holt-Giménez 2006). Farmer field schools have been shown to significantly reduce pesticide use, as chemical inputs are replaced by knowledge; large-scale studies in Indonesia, Vietnam, and Bangladesh recorded 35%–92% reductions in insecticide use in rice, and 34%–66% reductions in pesticide use in combination with 4%–14% better yields recorded in cotton production in China, India, and Pakistan (Berg and Jiggins 2007). Farmer field schools are also empowering farmers to better organize themselves, as well as stimulating continued learning.

An improved dissemination of knowledge by horizontal means transforms the nature of knowledge itself, making it the product of a network. It should encourage farmers, particularly small-scale farmers living in the most remote areas and on the most marginal land, to identify innovative solutions, working with experts toward a co-construction of knowledge that primarily benefits them, rather than only benefiting the better-off producers. This is key to enforcing the right to food. First, it enables public authorities to benefit from the experience and insight of farmers. Rather than treating smallholder farmers as beneficiaries of aid, they should be seen as experts with knowledge that complements formalized expertise. Second, participation can ensure that policies and programs are truly responsive to the needs of vulnerable groups, who will question projects that fail to improve their situation. Third, participation empowers the poor. And this is a vital step toward poverty alleviation: lack of power is a source of poverty, as marginal communities often receive less support than the groups that are better connected to government. Poverty in turn exacerbates this lack of power, creating a vicious circle of further disempowerment. The implication is that the organization of farmers in unions can effectively help overcome the collective-action problems that have made it so difficult for many small-scale farmers in the Global South to be involved in designing the policies that affect them; this is a key component of any scheme aimed at supporting them. Fourth, policies codesigned with farmers have a high degree of legitimacy and thus favor better uptake by other farmers. The participation of food-insecure groups in the policies that affect them should become a crucial element of all food security policies, from policy design, to the assessment of results, to decisions on research priorities. Indeed, improving the situation of millions of food-insecure peasants cannot be accomplished without them.

Works Cited

Ajayi, Olu C., and F. K. Akinnifesi. 2007. "Labor requirements and Profitability of Alternative Soil Fertility Replenishment Technologies in Zambia," 279–83. Accessed March 23, 2011. http://www.aaae-africa.org/proceedings2/005/Ajayi.pdf

Akande, Gbola and Y. Diei-Ouadi. 2010. "Post-harvest losses in small-scale fisheries: Case studies in five sub-Saharan African countries." FAO Fisheries Technical Paper no. 550.

Akinnifesi, Festus K. et al. 2010. "Fertilizer Trees for Sustainable Food Security in the Maize-based Production Systems of East and Southern Africa. A review." *Agronomy for Sustainable Development* 30 (3): 615–29.

Altieri, Miguel. 1995. *Agroecology: The Science of Sustainable Agriculture.* 2nd ed. Boulder: Westview Press.

Altieri, Miguel, and C. Nicholls. 2004. *Biodiversity and Pest Management in Agroecosystems*. 2nd ed. New York: Haworth Press.

Burney, Jennifer A. et al. 2010. "Greenhouse Gas Mitigation by Agricultural Intensification." *Proceedings of the National Academy of Sciences* 107 (26): 12052–57.

Campbell, B. et al. 1997. "Local Level Valuation of Savannah Resources: A Case Study from Zimbabwe." *Economic Botany* 51, 57–77.

Christiaensen, Luc, L. Demery, and J. Kuhl. 2011. "The (Evolving) Role of Agriculture in Poverty Reduction." *Journal of Development Economics.* Accessed March 23, 2011. http://www.wider.unu.edu/stc/repec/pdfs/wp2010/wp2010-36.pdf.

Christensen Fund, The, and Bioversity International. 2010. *The Use of Agrobiodiversity by Indigenous and Traditional Agricultural Communities in Adapting to Climate Change—Synthesis paper*. Platform for Agrobiodiversity Research. http://www.agrobiodiversityplatform.org/blog/wp-content/uploads/2010/05/PAR-Synthesis_low_FINAL.pdf.

Demment, Montegue W. et al. 2003. "Providing Micronutrients through Food Based Solutions: A Key to Human and National Development". *J. Nutrition* (133): 3879–85.

Diop, Amadou M. 2001. "Management of Organic Inputs to Increase Food Production in Senegal." In *Agroecological Innovations: Increasing Food Production with Participatory Development*, edited by N. Uphoff, 252–53. London:Earthscan.

Eyhord, Frank et al. 2007. "The Viability of Cotton-based Organic Agriculture Systems in India." *International Journal of Agricultural Sustainability* 5: 25–38.

FAO (Food and Agriculture Organization of the United Nations). 2006. "World Agriculture, Toward 2030/2050." Rome: FAO.

Frison, Emile, et al. 2006. "Agricultural Biodiversity, Nutrition and Health: Making a Difference to Hunger and Nutrition in the Developing World." *Food and Nutrition Bulletin* 27 (2): 167–79.

Garrity, Dennis P., et al. 2010. "Evergreen Agriculture: A Robust Approach to Sustainable Food Security in Africa." *Food Security* 2: 197–214.

Government Office for Science. 2011. Foresight. London. Accessed March 23, 2011. http://www.bis.gov.uk/assets/bispartners/foresight/docs/food-and-farming/11-547-future-of-food-and-farming-summary.pdf.

Hoffmann, Ulrich. 2009. "On the Mitigation Potential of Agriculture." UNCTAD presentation. Accessed March 23, 2011.http://vi.unctad.org/files/studytour/sttanzania10/files/week1/wednesday17feb/Present%20on%20UNCTAD%20TER%20UHoffman.ppt.

———. 2010. "Assuring Food Security in Developing Countries under the Challenges of Climate Change: Key Trade and Development Issues of a Profound Transformation of Agriculture." United Nations Conference on Trade and Development Discussion Paper no. 201, 5–6.

Holt-Giménez, Eric. 2002. "Measuring Farmers' Agroecological Resistance After Hurricane Mitch in Nicaragua: A Case Study in Participatory, Sustainable Land Management Impact Monitoring." *Agriculture, Ecosystems and the Environment* 93 (1–2): 87–105.

———. 2006. *Campesino a Campesino: Voices from Latin America's Farmer to Farmer Movement for Sustainable Agriculture*. Oakland: Food First Books.

ICRAF (World Agroforestry Centre). 2009. "Creating an Evergreen Agriculture in Africa for Food Security and Environmental Resilience." Nairobi. Accessed March 23, 2011. http://www.worldagroforestry.org/downloads/publications/pdfs/B09008.pdf.

IPCC (Intergovernmental Panel on Climate Change). 2007. *Climate Change 2007: Climate Change Impacts, Adaptation and Vulnerability*. Contribution of Working Group II to the Fourth Assessment Report of the Intergovernmental Panel on Climate Change. Cambridge and New York: Cambridge University Press.

Pretty, Jules. 2008. "Agricultural Sustainability: Concepts, Principles and Evidence." *Philosophical Transactions of the Royal Society B: Biological Sciences* 363 (1491): 447–65.

Pretty, Jules et al. 2006. "Resource-conserving Agriculture Increases Yields in Developing Countries." *Environmental Science and Technology* 40 (4): 1114–19.

Kassam, Amir, et al. 2009. "The Spread of Conservation Agriculture: Justification, Sustainability and Uptake." *International Journal Agricultural Sustainability* 7 (4): 292–320.

Kasterine, Alexander, and D. Vanzetti. 2010. "The Effectiveness, Efficiency and Equity of Market-based and Voluntary Measures to Mitigate Greenhouse Gas Emissions from the Agri-food Sector." *Trade and Environment Review* (2009/2010): 87–111. Geneva: UNCTAD.

Khan, Zeyaur, R. et al. 2011. "Push-Pull Technology: a Conservation Agriculture Approach for Integrated Management of Insect Pests, Weeds and Soil Health in Africa." *International Journal of Agricultural Sustainability* 9 (1).

Landers, John, N. 2007. "Tropical Crop-Livestock Systems in Conservation Agriculture: The Brazilian Experience." *Integrated Crop Management* 5. Rome: FAO.

Lipton, Michael. 1977. *Why People Stay Poor: A Study of Urban Bias in World Development*. London: Maurice Temple Smith.

McIntyre, Beverly D., Hans R. Herren, Judi Wakhungu, and Robert T. Watson, eds. 2009. *Agriculture at a Crossroads: The International Assessment of Agricultural Knowledge, Science and Technology for Development; Summary for Decision Makers of the Global Report.* Washington, DC: Island Press.

Mele, Paul van, et al., eds. 2005. "Integrated Rice-Duck: a New Farming System for Bangladesh." *Innovations in Rural Extension: Case Studies from Bangladesh*. Dhaka: CABI Publishing.

Pye-Smith, Charlie. 2008. *Farming Trees, Banishing Hunger. How an Agroforestry Program Is Helping Smallholders in Malawi to Grow More Food and Improve their Livelihoods.* Nairobi: World Agroforestry Centre (ICRAF). Accessed March 23, 2011. http://www.worldagroforestry.org/downloads/publications/PDFS/b15589.pdf.

———. 2010. *A Rural Revival in Tanzania: How Agroforestry Is Helping Farmers to Restore the Woodlands in Shinyanga Region*. Nairobi: World Agroforestry Centre (ICRAF). Accessed March 23, 2011. http://www.worldagroforestry.org/downloads/publications/pdfs/B16751.pdf.

Smith, P. et al. 2007. "Agriculture." In *Climate Change 2007: Mitigation of Climate Change*. Contribution of Working Group III to the Fourth Assessment Report of the Intergovernmental Panel on Climate Change. Geneva: IPCC.

Stern, Nicholas. 2007. *Stern Review: Report on the Economics of Climate Change*, 67–68. Cambridge, UK: Cambridge University Press.

UNCTAD (United Nations Conference on Trade and Development), and UNEP (United Nations Environmental Programme). 2008. "Organic Agriculture and Food Security in Africa." UNEP-UNCTAD Capacity Building Task Force on Trade, Environment and Development (UNCTAD/DITC/TED/2007/15), 16-17. New York and Geneva: United Nations.

UNDP (United Nations Development Programme). 2007. *Human Development Report 2007/2008. Fighting climate change: Human solidarity in a divided world*, 90–91. New York: UNDP.

UNEP (United Nations Environmental Programme). 2009. *The Environmental Food Crisis*, 27–28. Birkeland, Norway: Birkeland Trykkeri AS.

Van den Berg, H., and J. Jiggins. 2007. "Investing in Farmers: The Impacts of Farmer Field Schools in Relation to Integrated Pest Management." *World Development* 35 (4): 663–86.

World Bank. 2008. *Word Development Report 2008: Agriculture for Development*, 6–7. Washington D.C.: World Bank.

Zhu, Y. Y. et al. 2000. "Genetic Diversity and Disease Control in Rice." *Nature* 406: 718–22.

Chapter Fifteen

AGRICULTURE AT A CROSSROADS

By Hans R. Herren,
President Millennium Institute, with Angela Hilmi

Introduction

THE AG ASSESSMENT, as the International Assessment of Agricultural Knowledge, Science and Technology for Development (IAASTD) has also been called, is a unique event in the history of agriculture and in human history in general. Never before has there bteen such an orchestrated effort to bring together, from every corner of the planet, knowledge, thoughts, and experiences on the most deeply rooted practice of humankind, the cultivation and care of plants and animals to fulfill livelihood and cultural needs, and beyond.

The IAASTD was brought to the international agenda at a crucial time, when the international community, in the middle of the 2008 food crisis, realized that the human species is now at a crossroads and that, unless we choose to act quickly, seriously, and at scale, we run the very real risk of collapse. The IAASTD brought facts, figures, and scientific evidence about the food and nutrition security challenges we are faced with today. It provided clarity and the data required to draw conclusions and propose options for actions for the future.

The IAASTD also brought hope. It showed that if business as usual is not an option, there are other possible paths ahead to take, and that these are within reach, provided decisive political, institutional, and financial support is put in place.

Process

The scope as defined by a series of multi-stakeholder consultations around the world was wide, the means followed, brought by different organizations and countries, coming from diverse horizons.

The IAASTD goals were to assess the impacts of past, present, and future agricultural knowledge, science, and technology (AKST) on (a) the reduction of hunger and poverty; (b) the improvement of rural livelihoods and human health; and (c) equitable and socially, environmentally, and economically sustainable development.

The World Bank and the Food and Agriculture Organization of the United Nations (FAO) initiated the IAASTD in 2002. The question was to determine whether an international assessment of agricultural knowledge, science, and technology was needed. The answer was, indisputably, *yes*. The result was an astounding collection of knowledge, analysis, modeling, and information, with the latest state-of-the-art formal and informal science, on all agriculture-related issues. For the first time, a global assessment brought together a wealth of knowledge on interconnected major contemporary challenges. The IAASTD became the inescapable and essential reference for any practitioner and decision maker in this field. An additional, often undervalued outcome of the assessment was capacity development, which was meant to elicit similar assessments at the national level (since there were authors from most countries in the five regions) in order to focus the options for action at a more local level, as is eminently relevant in agriculture.

The outputs from the assessment were pulled together in one global and five subglobal reports; a global and five subglobal summaries for decision makers; and a crosscutting synthesis report with an executive summary. The decision makers' summaries and the synthesis report specifically provide options for action to governments, international agencies, academia, research organizations, and other decision makers around the world.

Hundreds of experts from all regions of the world participated in the preparation and peer-review process. It is the synergy of these interrelated disciplines that created the uniqueness of this interdisciplinary regional and global process. At the onset of the project, the concept was endorsed as multi-thematic, multi-spatial and multi-temporal. The intergovernmental process included a multi-stakeholder bureau cosponsored by FAO, the Global Environment Facility, United Nations Development Programme (UNDP), the United Nations Environmental Programme (UNEP), the United Nations Educational, Scientific and Cultural Organization (UNESCO), the World Bank, and the World Health Organization. The IAASTD's governance structure was a unique hybrid of the Intergovernmental Panel on Climate Change and the nongovernmental Millennium Ecosystem Assessment. A multi-stakeholder bureau comprising 30 representatives from government and 30 from civil society governed the assessment. The process brought together 110 governments and 400 experts, representing nongovernmental organizations, the private sector, producers, consumers, the scientific community, and multiple international agencies involved in the agricultural and rural development sectors.

Achim Steiner, executive director of UNEP, opened the final intergovernmental plenary in Johannesburg, South Africa, on April 7, 2008. An overwhelming majority of governments approved the summaries for decision makers and the executive summary of the synthesis report.

Key Findings and Options for Action

The IAASTD goals were consistent with major UN Millennium Development Goals: the reduction of hunger and poverty; the improvement of rural livelihoods and human health; and the facilitation of equitable and socially, environmentally, and economically sustainable development. Realizing these goals requires acknowledging the multifunctionality of agriculture, which means "the inescapable interconnectedness of agriculture's different roles and functions, whereby agriculture is understood as a multi-output activity producing not only commodities (food, feed, fibers, agrofuels, medicinal products and ornamentals), but also non-commodity outputs such as environmental services, landscape amenities, and cultural heritages" (McIntyre et al. 2009).

There was a general agreement that the scope of the assessment needed to go beyond the narrow confines of science and technology to encompass other types of relevant knowledge (e.g., knowledge held by agricultural producers, consumers, and end users), and that it should also assess the role of institutions, organizations, governance, markets, and trade. However, the inclusion of knowledge presented a major challenge in the management of the assessment, in particular during the review process, where many statements issued from the knowledge sphere were challenged by reviewers used to the rigor of peer-reviewed journals.

The IAASTD report, published under the title of *Agriculture at a Crossroads*, concluded that modern as well as traditional agriculture will have to change radically if the world is to avoid social breakdown and environmental collapse. Further, it suggests that "business as usual" is not an option and that change, or a new paradigm for agriculture, is inescapable, imperative, and urgent.

The report, with the authors' brief being to examine hunger, poverty, the environment, and equity together; is definitively pro-poor. In an interview about the IAASTD (Wilson 2008), its director, Professor Robert Watson, former chief scientist of the World Bank, says those on the margins are ill-served by the present system: "The incentives for science to address the issues that matter to the poor are weak . . . the poorest developing countries are net losers under most trade liberalization scenarios." The report's authors concluded that liberalizing agricultural trade did not help small farmers or rural communities appreciably in much of the world. Opening national agricultural markets to international competition before basic infrastructure and national institutions are in place undermines progress in agriculture, poverty alleviation, the environment, and food security.

Further, they also concluded that the willingness of many people to tackle the basics of linking production, social, and environmental goals is marred by "contentious political and economic stances" (McIntyre et al. 2009). More specifically, this refers to the many Organisation for Economic Co-operation and Development (OECD) member countries, which are deeply opposed to any changes in trade regimes or subsidy systems. Without major reforms in the industrialized countries, many poorer countries will have a very hard time moving toward an efficient and productive agricultural sector. There is also criticism of corporate, profit-oriented agriculture that is short term, exploitative of soil and water, and lacks the diversity that would promote resilience in the face of the climate change challenges as well as assure improved health. The report therefore calls for a "multifunctional" agriculture—serving our multiple demands for sustainability, equity, development, and food and nutrition. There are, it concludes, "unprecedented challenges ahead" (McIntyre et al. 2009). It expresses the belief that the way to meet them lies in redirecting the wealth of agricultural knowledge, science, and technology (AKST) the world has built up; this should be targeted toward agroecology strategies that combine productivity with protecting natural resources like soils, water, forests, and biodiversity. In particular, the research and development efforts must now target and include in a participatory manner small-scale and family farmers, since they make up the major part of the poor and hungry, while they also represent the major part of the stewards for the environment. Agricultural practices like organic, biodynamic, conservation, and agroecology are suggested as options that address the main constraints to food and nutrition security as well as food sovereignty issues.

Agriculture, with its basic sustainable development role, calls for new and redirected AKSTs to reverse the world's growing inequities. The report states: "AKST alone cannot solve these problems . . . but it can make a major contribution" (McIntyre et al. 2009). In many countries, it says,

> Food is taken for granted, and farmers are poorly rewarded and respected for putting it on our tables, let alone as stewards of almost a third of the Earth's land. Investment in agricultural science and its extension to farmers has decreased, yet we still urgently need sustainable, environmentally sound and equitable ways to produce food (McIntyre et al. 2009).

While looking at food systems, the IAASTD report, which covers agriculture in its widest sense, gives special attention to a select number of areas that present a major interest and are of concern to policy makers:

Bioenergy New efforts are needed to improve the traditional bioenergy (like wood fuels) on which millions of people depend. First-generation biofuels (mainly bioethanol and biodiesel) compete for land and water with food crops, and are uneconomic in most parts of the world. Their successors, such as cellulosic ethanol and biomass-to-liquids technologies, may reduce agricultural land requirements per unit of energy produced. Their environmental and social effects are, however, uncertain: they could, for instance, by using farm waste, deprive soils of essential organic matter and also compete for water and land.

Biotechnology There is a significant lack of transparent communication by those involved, so assessment of the technology lags behind its development, information can be anecdotal and contradictory, and uncertainty about possible benefits and risks is unavoidable. The use of genetically modified crops, where the technology is not contained, is contentious. Results from short-term experiments are yet to show a significant benefit, in particular at a small-scale farm level, and for major tropical staple crops. Also, there is little information from publicly funded research on long-term impacts on the environment and some key ecosystem services such as natural pest control. The report calls for more research on genetically modified crops before their deployment, a needs assessment and alternative options, while also promoting the use of molecular tools in classical breeding. Local priorities matter; biotechnologies should be used to maintain local expertise and germplasm, so that the capacity for further research rests in the community.

Climate Change Irreversible damage to the natural resource base on which agriculture depends is expected under even the most optimistic climate change scenarios. The relationship is a two-way street: agriculture adds significantly to climate change, and is generally adversely affected by it. While modest temperature rises may benefit higher latitudes by increasing yields, though not lower ones, further warming will negatively impact agriculture in all regions of the globe. The impacts of increased and unpredictable rain and droughts call for a resilient type of agriculture, in which soil can act as a buffer to these extreme conditions. As extreme weather becomes more frequent, there will also be serious potential for conflict over habitable land and resources like fresh water and land suitable for food, feed, and fiber production. The main research areas for AKST will be concerned with adaptation to climate change, while there will also be a need to work on the mitigation part by making agriculture less dependent on external energy in the form of fertilizers and agrochemicals.

Human Health In the past, improving human health has not generally been an explicit goal of agricultural policy. The authors of the IAASTD report, however, have addressed this concern of the assessment's stakeholders, by expanding food security to nutrition security. Agriculture needs to focus on consumers and the importance of dietary quality as main drivers of production, and not merely on quantity or price. Public health issues AKST could address include pesticide residues, heavy metals, hormones, antibiotics, food additives, and diversity of food sources. The health of agricultural workers is of equal concern; globally, the sector accounts for at least 170,000 occupational deaths annually, half of all fatal accidents. Other significant hazards include agrochemicals, transmissible animal diseases, toxins and allergens, and noise, vibration, and ergonomic problems. Many of the new infectious diseases now emerging thrive in intensified crop and livestock systems. Adding diversity to the food basket will have two major impacts: better nutrition and more productive and resilient cropping system.

Natural Resource Management Global agricultural development has focused narrowly on increased production, not on a more holistic regard for the sustainable use of natural resources—food and nutritional security together. Farmers need to be involved in a two-way learning process with policy makers, researchers, and civil society to shape natural resource management policies that benefit them and the consumers rather than the middlemen and input suppliers. When AKST is used creatively with the active participation of various stakeholders, the misuse of natural capital can be reversed and resources conserved, and eventually built up for future generations. A regenerative and multifunctional agriculture is what the authors of the report basically call for. One main statement was that animals need to find their way from factories back to the farms, thus facilitating the closing of the carbon cycle, the most basic tenet of sustainable agriculture, given its key role in rebuilding and maintaining soil fertility.

Trade and Markets Increasing equity for poor countries requires special and differential treatment and nonreciprocal access in place of across-the-board trade liberalization. Targeting market and trade policies to enhance the ability of AKST to promote development, food security, environmental sustainability, and the profitability of small farms is an immediate global challenge for reducing poverty and inequity. It is suggested that the production and other perverse subsidies be removed or redirected toward practices that promote sustainability; taxing the heavy externalities that come with industrial agriculture; producing better definitions of property rights; and developing rewards and markets for agroenvironmental services, including a carefully thought out extension of carbon financing to provide incentives for sustainable agriculture.

Traditional and Local Knowledge and Community-Based Innovation Formal, traditional, and local knowledge need to be integrated once AKST is directed at truly multifunctional agriculture—enhancing production, profitability, ecosystem services, and food systems. Information and communication technologies can help to achieve effective collaboration between scientists, researchers, and local and indigenous people. Examples of misappropriation of their knowledge and of community-based innovations show the need to share information about regulatory frameworks. The more formal research should be increasingly carried out in the public sector with public funding, to create new public goods that benefit society in general.

Women in Agriculture The involvement of women in the agricultural sector is growing in many developing countries, especially with the development of export-oriented irrigated farming. Most rural women worldwide face worsening health and work conditions, limited access to education and control over natural resources, insecurity in employment, and low income. Possible steps to remedy this include: giving priority to women's access to education, information, science and technology, and extension services; improving their access to, and ownership and control of economic and natural resources through laws, credit schemes, and support for income-generating activities; giving priority to women-farmer groups in value chains; supporting public services and investment in rural areas to improve women's lives; and assessing the effects of farming practices and technologies, including pesticides, on women's health, and reducing use and exposure. There is also the need to remove the hard work and drudgery from agriculture to reduce the burden farmers have to cope with, and also to keep the youth on the farm.

Regional Differences and Examples of Specific Emphasis in the Five Regions

As mentioned earlier, agriculture is a very local affair. The ecological and cultural conditions greatly affect all facets from production to consumption. It is therefore logical that the challenges and perception of relative importance of development and sustainability goals also vary accordingly. At the global, regional, and national levels, the IAASTD highlights that decision makers must be acutely conscious of the fact that there are diverse challenges, multiple theoretical frameworks and development models, and a wide range of options. Some examples of regional differences are provided below, to highlight these differences and make the point for even more refined assessments at the local level to assist decision makers with well-tailored options for action:

- *The commitment to address poverty and livelihoods* reflects the critical role of agriculture and rural employment opportunities in developing countries where 30%–60% of all livelihoods arise from agricultural and related activities. In North America and Europe (NAE), where food insecurity and hunger are no longer the major problems, attention has shifted to the question of relative poverty (McIntyre et al. 2009).
- *Reducing hunger* is an important goal in all developing regions: Central, West, and North Africa (CWANA), East and South Asia and Pacific (ESAP), Latin America and the Caribbean (LAC), and Sub-Saharan Africa (SSA). Of the 854 million malnourished people in 2001–2003, only 9 million were in the developed world; ESAP accounted for 61% of the total. In ESAP, however, this represents only 15% of the total regional population, while the 206 million malnourished SSA inhabitants represent 32% of the region's population (McIntyre et al. 2009).
- *Improving human health and nutrition* is critical for all regions. Malnutrition is a major cause of ill health and reduced productivity, particularly in SSA and CWANA. Food safety is an important health issue in all regions. Inappropriate application of AKST contributes to the increase in overweight, obesity, and chronic diseases that is being experienced in all countries.

- *Environmental goals* are important globally, with pressure on the environment due to relatively high industrialization, urbanization, and productivity-enhancing agricultural practices in NAE, and pressures to enhance productivity even at the cost of environmental goods and services in SSA.
- *Equity* is important across all regions. This goal draws attention to the current conditions of inequitable distribution and access to resources, and to overall income inequality, which is most extreme in LAC. Regional analyses (ESAP, LAC, and SSA) indicate that the unequal distribution of resources is a major constraint that shapes development needs and impedes the achievement of all other development and sustainability goals.

Next Steps

Following the findings of the IAASTD, new priorities have been defined and the complex dimension of agriculture is now better understood. Using the insight of the IAASTD, work is ongoing with system dynamics modeling based on the Millennium Institute's T21 model to represent agricultural development as a process involving many social, economic, and environmental factors, systematically applying the "what if?" questions to compare policies across scenarios. These models give, for example, the magnitude of economic and ecological advantages of more sustainable practices adopted, while oil and fossil fuel prices increase and chemical fertilizers become less competitive. A range of green simulations, undertaken for UNEP's Green Economy Report (Ayres et al. 2011), provide a clear picture of different strategies and their effects on natural resource stocks, greenhouse gases, employment, food production, and the investments needed to transition to the new agricultural paradigm that has been called for by the IAASTD report.

With these new tools in hand, it is now time to manage this transition process, and reorient direct research investments, in particular to support reduction of pre- and postharvest losses, and indirect ones (enabling conditions such as facilitation of market access, insurances, rural infrastructure, value chain, etc.) to support the transition to conservation and agroecology farming, and to reorient research and development to new forms of agriculture blending of the latest science with traditional knowledge and practices, and promoting innovation, creativity, and new forms of collaboration and decision making.

Conclusion

The IAASTD was a four-year collaborative effort, begun in 2004, that assessed our capacity to meet development and sustainability goals of reducing hunger and poverty, improving nutrition, health, and rural livelihoods, and facilitating social and environmental sustainability. Governed by a multi-stakeholder bureau comprising representatives from government and from civil society, the process brought together the whole range of actors ranging from governments to experts, NGOs, the private sector, producers, consumers, the scientific community, etc. In addition to assessing existing conditions and knowledge, the IAASTD used a simple set of model projections to look at the future, based on knowledge from past events and existing trends such as population growth, rural/urban food and poverty dynamics, loss of agricultural land, water availability, and climate change effects.

The IAASTD is crucial in the history of agricultural science assessments in that it assesses both formal science and technology, and local and traditional knowledge; addresses not only production and productivity but also the multifunctionality of agriculture; and recognizes that multiple perspectives exist regarding the role and nature of AKST. It was also unique in the sense that it was a capacity-development process, enabling the participants to get practice in undertaking assessments, a process different from writing scientific publications or reviews. The expectation was that countries would carry out national assessments, with the authors that participated in the IAASTD, to inform the development of new AKST policies in support of the needed transition to sustainable agricultural practices. This has not yet taken place and efforts are underway with a new project supported both by the Biovision Foundation and FAO to assist countries in that direction.

The uniqueness of the process is that, based on on-the-ground findings, it was able to ask a different question, a complex question: *How do we rethink our global food system so that it can nourish people, create healthy communities and economies, and sustain the planet over and beyond the old, usual leitmotif: how do we boost food production?*

That is in itself a total shift in paradigm, which has opened new avenues and different horizons for the international community to focus on. Its principal message is that *business as usual is not an option*. To achieve food security, we must place agriculture in its wider context. The future belongs to those who will be able to blend farmer knowledge and innovation with formal science. Only an agriculture that fosters rural economies and revives vibrant communities; that restores, not erodes, biological diversity and soil fertility; and builds resilient food systems will be able to take us forward.

The IAASTD demonstrates that the way forward to significantly reducing hunger is to enact policies and practices that ensure equitable access to food; reduce food waste and postharvest losses; rebuild lively local markets; and redirect the land and resources increasingly used to feed cars, animals, and industrial processes to nourish humans, while supporting smallholder farmers so they can maintain ownership of their own productive resources and be the stewards of the land that will sustain the generations to come.

Thanks to the data collected and to the quality of the analysis that brought a level of integration never attained before, the evidence in support of low-input, ecological, or conservation agriculture is now undeniable. And evidence that sustainable, biologically based agriculture can provide the nutrition and income to the billion-plus poor and hungry of today, and the two billion newcomers by 2050, is now well proven. How to operate this transition, how to truly support small farmers—the people who produce the majority of the world's food, safeguard our ecosystems, yet are the majority of the world's poor—is now the next step, opening a new course of action based on a different and renewed paradigm.

As a move to promote the use of the key findings and implement the options for actions of the IAASTD report, the Biovision Foundation is working with countries, civil society organizations, and intergovernmental agencies that participated in the assessment in assuring a prominent space at Rio+20 for sustainable and multifunctional agriculture as the cornerstone of sustainable development. It should be noted that it was in 2002 that the IAASTD was launched, at the UNCSD in Johannesburg; therefore Rio+20 is a great opportunity to reiterate the objectives of the assessment and also make the case for the implementation of its outcomes without delays.

Works Cited

Ayres, Robert, et al. 2011. "Green Economy Report." UNEP. St-Martin-Bellevue: 100 Watt.

McIntyre, Beverly D., Hans R. Herren, Judi Wakhungu, and Robert T. Watson, eds. 2009. "Agriculture at a Crossroads: The International Assessment of Agricultural Knowledge, Science and Technology for Development." Washington, DC: Island Press.

Wilson, Kelpie. 2008. "A Pre-Columbian Brazilian Practice Teaches World How to Grow Food." Brazil. Accessed May 21, 2011. http://www.brazzil.com/component/content/article/190-april-2008/10061-food-and-biofuels.html.

Chapter Sixteen

NOW'S THE TIME TO MAKE IT HAPPEN: THE UN COMMITTEE ON FOOD SECURITY

By Nora McKeon,
Coordinator, Terranuova

IT LOOKS LIKE it could be a magic moment for food activists! For once, three important ingredients for change in the global food system have come together simultaneously: vibrant, worldwide people's food movements building from the local level on up; cracks in the dominant, corporate-controlled "wisdom" about how best to ensure food for all; and a new and exciting global space for decision making on food issues. That's right, the United Nations (UN) system that many people had given up on as a tired and toothless bureaucracy is proving instead to be part of the solution, and an important one at that. Let's take a look at these three developments one by one.

First of all, people around the world are acting to take control of their food or, rather, to retake control. After all, locally centered food systems were a basic part of the texture of human society until three decades ago, when liberalization and globalization opened the door to a corporate takeover of what we grow and what we eat. Today, a rich and powerful range of alternatives to the corporate food system is emerging locally in all regions of the world.

This emergence has been happening among small farmers since the 1980s, especially in the Global South, in reaction to neoliberal policies promoted by the International Monetary Fund and the World Bank. Camouflaged by an inscrutable label—"structural adjustment"—that evoked something out of a chiropractor's manual, these measures arrived on the tail of the food crisis of the 1970s and provided the champions of the untrammeled free market with an opportunity to peddle their wares. State support to agriculture was scuttled. Agricultural credit, extension, price support, input provision, and marketing services all became things of the past. At the same time, developing-countries' markets were opened up to the 'fresh breeze of competition' from agricultural products originating in Europe and the US, where the 'harmful' practices from which the South was being weaned continued to be applied.

The effects on agricultural production and rural livelihoods were devastating. An analysis of the impact of structural adjustment in Senegal, conducted by the peasant movement Fédération des Organisations Non-Gouvernementales du Sénégal (FONGS) in 1993, reported that the state's abrupt retrenchment had resulted in a dramatic decline in access to credit and use of inputs. Rural revenues had dropped from 22,000 Central and West African francs (47 USD) in 1960 to 8,000 CFA (17 USD) in 1990 (McKeon, Wolford, and Watts 2004). This was when the peasant movement started to organize nationally in West Africa, as village-based associations grouped together to confront an unsustainable situation and occupy a space created by the withdrawal of the omnipresent state. The Landless Workers Movement (Movimento dos Trabalhadores Rurais Sem Terra; MST) saw the light of day in Brazil in the same period. The conditions there were different in many ways, but there, too, rural people were reacting to unacceptable conditions exacerbated by debt and structural adjustment, and were taking advantage of a moment of political opportunity. Similar developments were underway in other regions of the South, as well as in Europe and North America (Edelman 2003), where smallholder farmers constituted a far smaller proportion of the population but were subject to the same pressures as their Southern sisters and brothers.

The birth of the World Trade Organization (WTO) further encouraged networking among the primary victims of globalization and liberalization. The decision to set up La Via Campesina in 1993 was triggered by the Uruguay Round of the General Agreement on Tariffs and Trade (GATT), and the realization that agricultural policies would henceforth be determined globally and that it was essential for small farmers to be able to defend their interests at that level (McKeon and Kalafatic 2009; Desmarais 2007). The Network of Peasant and Agricultural Producer Organizations of West Africa (Réseau des Organizations Paysannes et de Producteurs de l'Afrique de l'Ouest; ROPPA) was established in 2000 with similar motivations. In the words of its first president, Ndiogou Fall, in 2002:

> The levels of decision-making seem to be jumping around like frogs. From the national scene, where our farmers' platform is well in place now, to the regional level, where we are managing to make our voice heard. But tomorrow it will be New Partnership for Africa's Development (NEPAD), the European Union, the World Trade Organization . . . The temptation of just buckling down to work in one's own field is strong. But it is no longer an option (McKeon 2009).

The civil society forums held in parallel to the two World Food Summits convened by the UN's Food and Agriculture Organization (FAO) in 1996 and 2002 gave a strong impetus to global networking by rural social movements that came to identify with the principle of food sovereignty. The organizers of these forums, unlike those of the NGO-dominated meetings that accompanied other UN summits, made sure that small food producers and indigenous peoples were in the majority by applying a quota system for delegates and mobilizing resources to cover their travel costs. The principle of food sovereignty was introduced by La Via Campesina at the 1996 forum. By 2002 it had become the assembly's battle cry. Its political statement, entitled "Food Sovereignty: A Right for All," was delivered to the official Summit by a Latin American peasant. The Forum mandated the International Planning Committee on Food Sovereignty (IPC) to carry forward the action agenda it adopted, based on four pillars (later expanded to six): the right to food and food sovereignty; mainstreaming agroecological family farming; defending people's access to and control of natural resources; and trade and food sovereignty. The IPC is an autonomous, self-managed global network of some 45 people's movements and NGOs involved with at least 800 organizations throughout the world. Its membership includes constituency focal points (organizations representing small farmers, fisher folk, pastoralists, indigenous peoples, agricultural workers), regional focal points, and thematic focal points (NGO networks with particular expertise on priority issues). It is not a centralized structure and does not claim to represent its members. It is rather a space for self-selected civil society organizations (CSOs) that identify with the food sovereignty agenda adopted at the 2002 forum.

Five years later, in February 2007, an important global encounter on food sovereignty held in Mali brought together more than 500 delegates from local movements and struggles in all regions; and deepened the common understanding of what food sovereignty means, what to fight for, and what to oppose. The Nyéléni Forum was organized by La Via Campesina, the World March of Women, the World Forum of Fish Harvesters and Fish Workers, the World Forum of Fisher Peoples, Friends of the Earth International, the International Planning Committee for Food Sovereignty, the Food Sovereignty Network, the Network of Peasant and Agricultural Producers Organizations of West Africa (the English-language name for ROPPA), and the National Coordination of Peasant Organizations of Mali.

Since then, the food sovereignty movement has continued to spread, not only in the Global South but also in the North, as communities have woken up to the impacts of the corporate-controlled food system. In the US, the annual conferences of the Community Food Security Coalition (CFSC) testify to the rich variety of local initiatives underway coast to coast: municipal food councils, food banks, urban agriculture, community stores, and more. The CFSC conference held in New Orleans in October 2010, witnessed the birth of a US Food Sovereignty Alliance (US Food Sovereignty Alliance, 2011), a step echoed in Canada just a month later (Food Secure Canada, 2011). For its part, Europe is rife with community-supported agriculture initiatives, public procurement of local food by municipalities for schools and hospitals, local seed trading fairs, farmers' markets, and regions that bind together in opposition to the introduction of genetically modified organisms (GMOs). A process is now underway to bring these local initiatives together in a horizontal, Europe-wide food sovereignty movement (Nyéléni 2011). Food sovereignty networks in the North are important not only to promote domestic community-based movements but also to target the power centers and policies that wreak havoc in the Global South. The growth of webs of smallholder food producers and community-based food initiatives over the past few years is fundamental, since the energy, inventiveness, know-how, and self-determination they express is ,without doubt, the indispensable basis for any successful effort to change food systems.

Six Pillars of Food Sovereignty:

1. **Focuses on Food for People**, putting the right to food at the center of food, agriculture, livestock, and fisheries policies; *and rejects* the proposition that food is just another commodity or component for international agribusiness.
2. **Values Food Providers** and respects their rights; *and rejects* those policies, actions, and programs that undervalue them, threaten their livelihoods and eliminate them.
3. **Localizes Food Systems**, bringing food providers and consumers closer together; *and rejects* governance structures,

agreements, and practices that depend on and promote unsustainable and inequitable international trade, and give power to remote and unaccountable corporations.

4. **Localizes Control** over territory, land, grazing, water, seeds, livestock, and fish populations; *and rejects* the privatization of natural resources through laws, commercial contracts, and intellectual property rights regimes.
5. **Builds Knowledge and Skills** that conserve, develop, and manage localized food production and harvesting systems; *and rejects* technologies that undermine, threaten, or contaminate these, e.g. genetic engineering.
6. **Works with Nature** in diverse, agroecological production and harvesting methods that maximize ecosystem functions and improve resilience and adaptation, especially in the face of climate change; *and rejects* energy-intensive industrialized methods that damage the environment and contribute to global warming.

(Source: Nyéléni 2007)

The second development supporting change is that while people's movements have been building up their strength, cracks have developed in the dominant global system they are opposing. Since late 2007, the food-price crisis and social unrest in cities around the world have thrown into question the failed food security strategies applied up to now. On the one hand, there has been a dethroning of the policy advice that transformed Africa from a net food exporter to a net food importer in the space of a decade. African countries were urged by the World Bank and the International Monetary Fund to exploit their "comparative advantage" by selling their raw commodities on the world market and buying "cheap" food, which producers from other regions were able to offer at below production costs thanks to the subsidies they received. The inanity of this approach became evident when food prices spiked three years ago and the low-income, food import–

dependent countries found themselves out on a limb. Even George Bush, hardly known for his radical and perspicacious views, had pointed out several years earlier: "Can you all imagine a country that isn't capable of growing enough crops to feed it? It would be a country exposed to international pressures, a vulnerable nation" (Suppan 2003).

Now there is fairly universal recognition of the need to support food production for domestic consumption by smallholder family farmers, which are the majority of the food insecure but also provide most of the food consumed in the South. Evidence that smallholder agroecological farming is capable of meeting the food needs of the world's population is mounting. The report of the International Assessment of Agricultural Knowledge, Science and Technology (IAASTD) calls for a fundamental paradigm shift in agricultural development, and advocates strengthening agroecological science and practice. Published in 2009, this report was the result of a four-year process involving 400 experts from all regions, sponsored by the FAO, and the United Nations Development Programme (UNDP) and the World Bank (McIntyre et al. 2009). The research of respected academics has certified that average yield increases of 79% can be obtained in sites around the world simply by adopting low-input, resource-conserving technologies (Pretty et al. 2006). The FAO is currently mobilizing the know-how of all of its technical divisions to publish a major work on the ecosystem approach to enhancing crop production, whereas just eight years ago the handful of agroecology advocates in the house practically had to slink around together in surreptitious corridor conversations to exchange ideas.

Simultaneously, the unsustainability of a food system based on intensive use of petrol products and chemical inputs has been dramatically highlighted by climate change and the energy crisis. According to recent United Nations Environmental Programme (UNEP) publications, the conventional agriculture model that prevails in the global food chain (strongly subsidized by both the European Common Agricultural Policy and the US Farm Bill) accounts for 14% of total annual greenhouse gas emissions. Most of this is due to the use of nitrogen fertilizers derived from rarifying petrol. Yet the agricultural sector could be largely carbon neutral by 2030 and produce enough food for a growing population, if localized agroecologically based systems were widely adopted (UNEP 2010). These practices have been proven to reduce emissions but currently lack policy and program support. In 2011, the Special Rapporteur on the Right to Food dedicated his annual report to the UN Human Rights Council to the topic. Entitled "Agroecology and the Right to Food," the report concludes that policies that support agroecology can contribute to adaptation to, and mitigation of, climate change, while simultaneously increasing yields and incomes in rural areas, and stimulating rural economies (De Schutter 2011). Production is not the only link in the chain affected; the entire globalized distribution process of the corporate-controlled world food system depends on its ability to discount the energy and petrol costs of whisking food around the world before it ends up on a supermarket shelf.

Dominant food security strategies have been cast into doubt in member countries of the Organisation for Economic Co-operation and Development (OECD) as well as in the South. Growing problems of obesity and unsafe food are drawing the attention of policy makers and the public at large to the fact that food system malfunctions impact the North as well. More people globally suffer from being overweight and obese than from hunger, and type 2 diabetes kills some 3.8 million people a year (Lang, Barling, and Caraher 2009; Nestle 2007). Type 2 diabetes is the sixth-leading cause of death in the US, affecting 8% of the population. Mad cow disease in the UK, salmonella in US eggs, and dioxin-affected Belgian chickens are just some of the recent examples of the food risks engendered by insufficiently and inappropriately regulated industrial food production and processing systems in the North. The link between food and health has been underscored in recent years, building alliances across previously segregated policy sectors.

One result of all of this crisis-stimulated rethinking has been a new openness toward concepts considered taboo or laughable over the past couple of decades, when neoliberal ideology and high-tech paradigms reigned supreme. Ideas like the right of Southern countries to protect their markets, as the US and Europe have been doing for years, or the notion that agroecology might be more than a hobby for quirky Berkeley professors and their hippie cohorts are now being seriously entertained in the debate about how to sustainably feed a growing population in a scenario of climate change.

So, first of all, we have people taking action around the world and articulating their opposition to the corporate food system, accompanied, secondly, by cracks in the paradigm that open up wiggle space for alternative approaches. And finally, we come to the third ingredient for change: for the first time in history, the international community has established a global policy forum for food issues where people's movements can defend their proposals. This may seem fairly remote from local action, but it is important because many factors that impact food systems escape the control not only of communities but even of national governments (just think of the "intellectual property rights" that allow corporations like Monsanto to sue organic farmers whose fields have been unintentionally contaminated with a neighbor's genetically modified canola).

Addressing these factors is precisely why La Via Campesina and other rural social movements took the trouble to come to the World Food Summits in Rome and mandated the IPC network to carry forward their Food Sovereignty Declaration and Action Plan. These movements made a strategic assessment that the FAO, as the UN system's "Ministry of Agriculture," could constitute a politically interesting intergovernmental policy forum alternative to the Bretton Woods institutions and the WTO. There were several reasons for this: more democratic governance with universal membership and a one country—one vote decision-making process; specific focus on food and agriculture and a mission to eliminate hunger; a mandate that includes a strong normative role and relative openness to engagement with civil society and rural people's organizations. Based on this assessment, the IPC and its rural social movement members have invested considerable energy in opening up meaningful political space within the FAO. Strongly rooted in rural and community movements in all regions, the IPC has combined the political legitimacy and mobilization capacity of people's organizations with the analytic and advocacy skills of NGOs in a mutually reinforcing relationship. Since 2003 it has facilitated the participation of over 2,000 representatives of small food producers' organizations in FAO policy forums where they had never set foot before, championing the right to food, food sovereignty, and agroecological food production as an alternative paradigm to free trade and green revolution technology (McKeon 2009; McKeon and Kalafatic 2009). This global policy space, and almost a decade of experience in occupying it, was ready to be exploited by the food sovereignty movement when the food crisis hit the headlines in late 2007.

The food-price crisis revealed an apparent global policy vacuum. In the absence of an authoritative and inclusive global body deliberating on food issues, decisions in this vital field were being made by international institutions like the WTO and the World Bank, for whom food security is hardly core business; by the economically powerful members of the G8/G20; and by transnational corporations and financial speculators subject to no political oversight. When the crisis erupted, a sharp divide emerged over how to fill the governance gap. On one side, the G8 threw up a veritable smoke screen of rhetoric about an elusive Global Partnership on Agriculture, Food Security, and Nutrition (GPAFS), promising billions of dollars of new investment in agriculture (that somehow hasn't quite materialized), and ever more advanced technological fixes for whatever ails society. An audacious alternative to the GPAFS was championed by a number of predominantly Southern governments allied with civil society organizations and social movements. Their plan aimed to transform the Committee on World Food Security (CFS), based in the FAO, from an ineffectual talk-shop into an authoritative, inclusive UN forum deliberating on food security in the name of ensuring the global right to food (Civil Society for the Committee on World Food Security 2011). The challenge was to effectively fill the global governance gap rather than simply paper it over and allow wealthy states and corporations to stay in control. Better the UN than the G8/G20, if it can actually be made to carry out its mission effectively.

This is what the CFS reform, which got underway in April 2009, was all about. The CFS chairperson, Maria del Carmen Squeff (Argentina's Permanent Representative to the FAO), led the process with passion and sagacity, taking the unusual step of opening it up to all governments and concerned stakeholders, including civil society. Organizations of Southern, smallholder food producers, assisted by the IPC, and NGOs made a fundamental contribution, interacting with governments on an equal basis. In the end, despite their diversity, the majority of participants came to feel a sense of ownership over the reform proposal that was adopted by acclamation during the 35th Session of the FAO CFS on October 17, 2009. As the head of a key delegation put it in a private conversation, "When this whole exercise got underway we were far from optimistic. We felt the CFS was a dead duck. Now it may not be a swan yet, but it certainly is up in the air and flying." The final reform document includes some very important points that civil society fought hard to defend against the attacks of a few governments that wanted to keep the new CFS as toothless as possible.

On paper, there has never been anything even remotely like this document in the firmament of global food governance. And the first session of the new CFS, in mid-October 2010, proved that this forum can make a difference in practice as well. The inaugural meeting was preceded by a two-day consultation in which civil society delegates prepared their positions and endorsed the mechanism for relating to the CFS, which had been autonomously designed in consultation with networks around the world (CFS 2011). The CFS agenda—which civil society had helped define—included some hot policy issues. One of these was how to address food-price volatility, the market dysfunction that sparked the 2007 food riots and is expected to stay with

The Reform Document of the Committee on World Food Security: Some Important Features

- Recognizes the structural nature of the causes of the food crisis and acknowledges that the primary victims are smallholder food producers.
- Defines the CFS as "the foremost inclusive international and intergovernmental platform" for food security, based in the UN system."
- Explicitly includes defending the right to adequate food in the CFS's mission.
- Recognizes civil society organizations—small food producers and urban movements especially—as full participants, for the first time in UN history. Authorizes them to intervene in debate on the same footing as governments and affirms their right to autonomously self-organize to relate to the CFS.
- Enjoins the CFS to negotiate and adopt a Global Strategic Framework (GSF) for a food strategy providing guidance for national food security action plans as well as agricultural investment and trade regulations. Evidence of the effectiveness of alternative approaches brought to the CFS by social movements and civil society organizations will feed into the GFS.
- Empowers the CFS to make decisions on key food policy issues, and promotes accountability by governments and other actors.
- Arranges for CFS policy work to be supported by a high-level panel of experts in which the expertise of farmers, indigenous peoples, and practitioners is acknowledged alongside that of academics and researchers.
- Recognizes the principle of "subsidiarity" (decisions should be made at the lowest appropriate level). Strong linkages to be built between the global meetings of the CFS and regional and country levels. Governments have committed to establishing national multi-stakeholder policy spaces in the image of the global CFS open to social movements and civil society organizations.

Source: FAO 2009.

us throughout the foreseeable future. The free trade defenders tried to limit the discussion to putting more effective safety nets in place to attenuate the impact of volatility on vulnerable sectors of the population. The civil society delegates and allied governments, on the contrary, fought to open the discussion up to include finding solutions to the causes of volatility—including financial speculation. The latter line won. The next session of the CFS, in October 2011, was asked to adopt a comprehensive proposal on which the food sovereignty movement's ideas about how to curb speculation, regulate markets, and guarantee remunerative prices for small producers will be brought to bear. In the meantime, the G20 has recognized the nascent authority of this new and more inclusive policy forum by announcing that the CFS's policy decisions will influence the outcome of the G20's own discussions on price volatility later in 2011.

A second explosive issue had to do with the outrageous phenomenon, sparked by the food crisis, that has come to be labeled "land grabbing." Here, too, there was a strong confrontation between two positions. Some of the G8 powers took the line that the surge in large-scale foreign investment in developing-country agriculture, including land acquisitions, was to be welcomed as a major contribution to solving the food crisis by producing more food and stimulating the economy. All that was needed was to "discipline" it with a code of conduct—the Responsible Agricultural Investment (RAI) Principles, formulated in closed-door discussions by the World Bank and other multilateral institutions—which investors could be asked to voluntarily apply to their operations.

On the other side, the IPC network and other civil society organizations denounced the RAI Principles as a move to legitimize the corporate takeover of rural people's territories, and pooh-poohed the idea of rational and virtuous corporate self-regulation: "Deciding who has rights over which land resources is essentially a political matter that involves conflicting interests and power relations. The RAI initiative's framework of land and resource rights focuses on technical issues; it is essentially blind to politics" (Global Campaign on Agrarian Reform 2010). They supported a different route to global rules making: guidelines on land tenure governance that the FAO has been developing over the past two years, in broad consultation with governments and civil society in all regions. The hard-fought final decision went in favor of the civil society movement position. The RAI Principles were not endorsed by the CFS. On the contrary, it was agreed that the FAO land tenure guidelines will be put to the next CFS session for negotiation and adoption as a political commitment by FAO member governments. Civil society participants at the coming CFS session will also be advocating a moratorium on land grabbing until the new guidelines have been put in place. "The Dakar Declaration on Land Grabbing" adopted at the February 2011 World Social Forum (WSF) specifically refers to the CFS and urges adoption of the FAO guidelines. This rare reference to a UN agency in WSF literature is a demonstration of the thesis advanced at the beginning of this article: the UN system—or at least this part of it—is well on its way to becoming an important part of the solution.

The link between social movement, global policy advocacy, and local struggles is fundamental. Applied to the land-grab issue, the rapidity with which land deals are proceeding adds even greater urgency. In the words of Ibrahima Coulibaly, president of the National Peasant Platform in Mali, where a Libyan company has occupied 100,000 ha of highly productive irrigated land in the rice-producing region, with the complicity of national capital and authorities, "The only global-level action that might make a difference in the immediate future would be if the CFS adopted a moratorium on land grabbing and mandated a mission to verify the situation" (Coulibaly 2010). Immediate global intervention may be necessary to halt extreme violations of human rights, but in the longer term, the only guarantee for people's rights lies in promoting accountable national governance. As a national workshop on land-tenure governance in Senegal, held in December 2010, noted, the land-grab phenomenon has activated the interest of a range of official and civil society actors in land-tenure questions, and created an opportunity to initiate a real dialogue between the state, producers' organizations, local authorities, and other partners. This, however, can only be possible if the state undertakes to facilitate a policy dialogue framework in which a meaningful debate and negotiation can take place on the objectives and concrete modalities of a land-tenure reform, something it has resisted up to now (IPAR et al. 2010). An important piece of the CFS reform is the commitment on the part of member governments to replicate at national and regional levels the multi-stakeholder approach that has been institutionalized in the global CFS. This commitment is an instrument that social movements and civil society organizations can use to push for transparency and accountability on the part of national governments. Conversely, participation in global policy discussions by governments that are held accountable to their citizens is the most important contribution one can imagine to attaining effective and equitable global governance.

As the October 2010 CFS session came to a close, the head of a delegation not particularly enraptured with the positions that civil society participants were championing, took the floor to state his view that the strong presence of civil society in the renewed committee had proved to be the most important aspect of the reform. "They call our bluff and say it like it is. We need them in the room." There is certainly no call for complacency. The corporate food system is poised to turn to its advantage whatever crisis comes around the corner, and financial speculators won't drop the food-commodity bone unless some kind of Pavlovian mechanism is installed. But we do have a fairly exceptional political opportunity at hand. There are cracks in the corporate armor, people's food sovereignty movements have never been stronger, and there's a new global forum in which their experience and proposals can be brought to bear. Small food producers and civil society organizations have played a decisive role in opening up this space. Now let's make it work to our advantage!

Works Cited

Civil Society for the Committee on World Food Security. 2011. Accessed February 20, 2011. http://cso4cfs.org.

Coulibaly, Ibrahima. 2010. Personal interview.

De Schutter, Olivier. 2011. "Agroecology and the right to food." Report of the Special Rapporteur on the Right to Food to the Human Rights Council. Geneva: Geneva Human Rights Council. A/HRC/16/49. Geneva.

Desmarais, Annette Aurélie. 2007. *La Via Campesina: Globalization and the Power of Peasants.* Halifax: Fernwood Publishing.

Edelman, Marc. 2003. "Transnational Peasant and Farmer Movements and Networks." In *Global Civil Society 2003*, 185–220. London: Center for the Study of Global Governance, London School of Economics and Political Science.

FAO (Food and Agriculture Organization of the United Nations). 2009. "Reform of the Committee on World Food Security. Final Version." CFS: 2009/2 Rev.2. Rome: FAO Committee on World Food Security.

Food Secure Canada. 2011. Home page. Accessed February 25, 2011. http://foodsecurecanada.org.

Global Campaign for Agrarian Reform. 2010. "Why We Oppose the Principles for Responsible Agriculture Investment." Accessed February 20, 2011. http://www.fian.org.

IPAR (Initiative Prospective Agricole et Rurale) et al. 2010. *Gestion foncière au Sénégal: Enjeux, état des lieux et débats. Actes de l'atelier.* Dakar: IPAR, 11.

Lang, Tim, D. Barling, and M. Caraher. 2009. *Food Policy. Integrating Health, Environment and Society.* Oxford: Oxford University Press, 112.

McIntyre, Beverly D., H. R. Herren, J. Wakhungu, and R. T. Watson, eds. 2009. *Agriculture at a Crossroads. The International Assessment of Agricultural Knowledge, Science and Technology for Development.* Washington, DC: Island Press.

McKeon, Nora. 2009. *The United Nations and Civil Society. Legitimating Global Governance—Whose Voice?* London and New York: Zed, xii; 50–120.

McKeon, Nora, and C. Kalafatic. 2009. *Strengthening Dialogue: UN Experience with Small Farmer Organizations and Indigenous Peoples.* New York: United Nations NGO Liaison Service, 3; 17–18.

McKeon, Nora, and W. Wolford, and Michael Watts. 2004. *Peasant Associations in Theory and Practice.* Geneva: United Nations Research Institute for Social Development, 14; 29.

Nestle, Marion. 2007. *Food Politics: How the Food Industry Influences Nutrition and Health.* University of California Press, 7.

Nyéléni. 2007. Forum for Food Sovereignty, 76. Accessed June 6, 2011. http://www.foei.org/en/resources/publications/food-sovereignty/2000-2007/nyeleni-forum-for-food-sovereignty/view.

Nyéléni. 2011. Nyéléni 2007. Forum for Food Sovereignty, 76. Accessed February 20, 2011. http://www.nyelenieurope.net.

Pretty, Jules et al. 2006. "Resource-conserving agriculture increases yields in developing countries." *Environmental Science and Technology*, 40(4) (2006): 1114–19.

Suppan, S. 2003. "Food Sovereignty in an Era of Trade Liberalisation: Are Multilateral Means Towards Food Sovereignty Feasible?" *Global Security and Cooperation Quarterly*, 9, Program on Global Security and Cooperation.

UNEP (United Nations Environmental Programme). 2010. *Agriculture: A catalyst for transitioning to a green economy.* A UNEP Brief. Nairobi: UNEP.

US Food Sovereignty Alliance. 2011. Accessed February 20, 2011. http://www.usfoodsovereigntyalliance.org.

Chapter Seventeen

FOOD SOVEREIGNTY AND CLIMATE JUSTICE

By Brian Tokar,
Institute for Social Ecology

Climate change is already seriously impacting us. It brings floods, droughts and the outbreak of pests that are all causing harvest failures. I must point out that these harvest failures are something that the farmers did not create. Instead, it is the polluters who caused the emissions who destroy the natural cycles . . . [W]e will not pay for their mistakes.

Henry Saragih, General Coordinator of La Via Campesina, Copenhagen, December 2009.

The immense challenge humanity faces of stopping global warming and cooling the planet can only be achieved through a profound shift in agricultural practices toward the sustainable model of production used by indigenous and rural farming peoples, as well as other ancestral models and practices that contribute to solving the problem of agriculture and food sovereignty.

"People's Agreement," World People's Conference on Climate Change and the Rights of Mother Earth, Cochabamba, April 2010

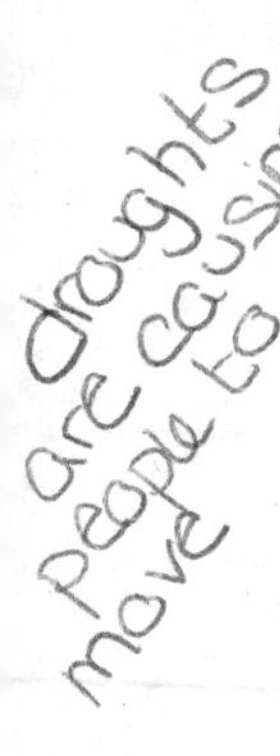

In the summer of 2010, nearly 20 million residents of Pakistan's Indus River valley—a fifth of the country's population—were forced to flee their homes and fields as the most severe monsoon rains in nearly a century buried much of the region under water. Now, areas of Nigeria and Cameroon are facing a different kind of deluge—a human wave of climate migrants, forced to flee their homes in the vicinity of Lake Chad, which has lost 90% of its area in recent decades due to a relentless long-term drought (Ngalme 2010). Farmers in the once Fertile Crescent of Syria and Iraq—the center of origin for wheat and barley, and perhaps agriculture as we know it—are similarly fighting a losing battle against an ever-encroaching desert (Worth 2010).

Most climate scientists will correctly point out that local phenomena and particular weather events cannot be linked unambiguously to the growing destabilization of the earth's climate systems. At the same time, long-term trends have convinced many that climate changes are happening much faster than even the best analytical models once predicted. After less than one degree Celsius of global warming, people around the world are experiencing increasingly chaotic weather patterns, including extreme heat waves and unprecedented cycles of flooding and drought. Barring an extraordinarily rapid change in our energy use and economic systems, the world will undoubtedly see several additional degrees of warming before the end of this century, with catastrophic consequences for many vulnerable people and ecosystems.

It is clear that some people will be far more affected than others, and that indigenous peoples and subsistence farmers in the tropics and subtropics are already experiencing the most serious consequences of an unfolding global climate crisis. This profound inequity in the impacts of climate changes, and the need for a thoroughgoing social transformation in order to alleviate it, are the primary insights of an emerging global movement known as "climate justice."

Climate justice advocates believe that the unfolding global climate crisis demands an unprecedented convergence of social movements. And there is no question that farmers and others concerned about the future of our food are central to this convergence. Today's dominant agricultural practices contribute greatly to changes in the climate, even as climate disruptions threaten people's ability to sustain traditional ties to the land and to survive. The future of traditional and sustainable agricultures significantly hinges on our ability to fend off catastrophic climate changes, and accomplishing this, in turn, requires that we overturn our civilization's dominant patterns of energy and land use, economic activity, and sociopolitical organization. The worldwide call for food sovereignty—the fundamental right of all people to control the means of raising their food—has become a staple of the climate justice agenda, and the climate crisis is further radicalizing the outlook of food activists throughout the world.

The concept of climate justice reflects somewhat different origins and emphases in various parts of the world. The term was first articulated by Indigenous Environmental Network founder and director Tom Goldtooth in the mid-1990s, was further defined in a 1999 Corpwatch report (Bruno, Karliner, and Brotsky 1999), and formed the basis for a resolution passed at the Second National People of Color Environmental Leadership Summit in the US in 2002. The concept gained international attention following a meeting in Durban, South Africa, in the fall of 2004, that included representatives of social movements and indigenous peoples' organizations based in Brazil, India, Samoa, the US, and the UK, as well as South Africa. That gathering crystallized around the drafting of the first comprehensive international declaration to challenge the emerging global carbon market, which is viewed by climate justice activists as a corporate-driven attempt to commodify the atmosphere and perpetuate and rationalize, rather than curtail, the expansion of fossil fuel use (Durban Group 2004). Carbon markets have failed to reduce emissions while offering new indirect subsidies to polluting industries, and thus are increasingly viewed as a false solution to the global climate crisis (Tokar 2010; Lohmann 2006).

In December 2007, representatives of peoples immediately affected by climate change demonstrated a strong presence at that year's UN climate summit in Bali, leading to a more formal worldwide network that emerged under the slogan "Climate Justice Now!" At subsequent UN conferences, indigenous peoples' and other land-based movements from the Global South, have reinforced their demands for climate justice. Some 20,000 representatives from popular movements around the world converged in Cochabamba, Bolivia, in April 2010, for a weeklong climate justice gathering that was convened by President Evo Morales, following the disappointing outcome of the UN's Copenhagen climate summit the previous winter. The World Conference on Climate Change and the Rights of Mother Earth, as it was called, produced a comprehensive platform and "People's Agreement" (World People's Conference on Climate Change and the Rights of Mother Earth 2010). Participants aimed to bring the results of those proceedings into the formal UN climate negotiations in Cancún and beyond.

In the United States, the demand for climate justice is voiced most articulately by environmental justice activists, mainly from communities of color that have been resisting daily exposure to chemical toxins and other environmental hazards for the past 30 years or more. At an important 2009 conference organized by West Harlem Environmental Action (WE ACT) in New York City, speakers described the emerging climate justice movement as a continuation of the civil rights legacy, and of the continuing "quest for fairness, equity and justice," as described by the pioneering environmental justice researcher and author, Robert Bullard (2009).

In Europe, climate justice has become a rallying cry for those who want to focus systemically on the institutions and corporations responsible for the continuing impasse in international climate negotiations. The European Climate Justice Action network's 2009 declaration stated, "We cannot trust the market with our future, nor put our faith in unsafe, unproven and unsustainable technologies. Contrary to those who put their faith in 'green capitalism,' we know that it is impossible to have infinite growth on a finite planet" (Climate Justice Action 2009). A discussion paper drafted in February 2010 explicitly connected food and agriculture to the call for climate justice:

> Climate Justice is closely linked to breaking the circle of industrialised agricultural production perpetuated through WTO and European policies. Speculation on food as an industrial commodity and the domination of long unsustainable production chains by international capital threaten the biosphere and the lives of billions of people. This attack on food sovereignty and the planet must be met with a social struggle for food production defined by the needs and rights of local communities. This means redefining, re-localising and re-appropriating the control of our food and agricultural systems through engaging and acting in solidarity with existing struggles (Climate Justice Action 2010).

Food and Climate: A "Two-Way Street"

While climate disruptions are already having profound effects on those who grow our food, the practices of industrial agriculture are largely responsible for altering the climate. The International Assessment of Agricultural Knowledge, Science and Technology for Development (IAASTD), a collaborative effort by four UN agencies and the World Bank, affirmed in a 2009 report that "the relationship between climate change and agriculture is a two-way street; agriculture contributes to climate change in several major ways and climate change in general adversely affects agriculture" (McIntyre et al. 2009, 8).

The IAASTD report acknowledged several well-known consequences of climate changes for agriculture, especially the increasingly disruptive effects on water cycles, and highlighted sustainable agricultural practices as a primary mitigation strategy. The report was widely acclaimed for its acknowledgment that traditional and local agricultural knowledge is central to the attainment of worldwide sustainability and development goals (McIntyre et al., 11). As several recent studies have noted, peasants and other smallholders contribute far more to the world's agricultural output than is widely acknowledged, and small farms are generally far more productive relative to their size than larger, industrial-scale farms (ETC Group 2009; Rosset 2010).

Estimates of the current food system's contribution to global emissions of greenhouse gases vary widely, from 10%–20% at the low end, up to nearly 60% (Wightman 2006; Saragih 2009). One anomalous but widely reported study suggests that livestock alone may be responsible for 51% of global emissions (Goodland and Anhang 2009). This range of estimates is the result of widely varying assumptions about key factors such as animal nutrition and waste handling, land and soil management practices, the impacts of the processing and transportation of food, and agriculture's contribution to global deforestation. These practices vary widely, of course, with the scale and cultivation methods of various-sized farms. The calculations are also complicated by the fact that greenhouse gases such as methane and nitrous oxide (N_2O) have far greater climate-forcing potential than carbon dioxide in the short term, by factors of 25 and 300, respectively, but do not remain in the atmosphere for nearly as long as CO_2 does.

In the United States, where official data suggest that only 8% of our emissions result from agriculture, the leading contributors are N_2O—produced when soil bacteria digest chemical-fertilizer residues—and methane released by ruminants, along with the CO_2 directly released from fossil fuel consumption (Paustian et al. 2006, 2). These three factors represent more than 80% of US agriculture's total emissions, and are highly sensitive to changes in livestock feed, manure management, fertilizer use, and the use of farm machinery.

Researchers who support the widespread adoption of organic agricultural methods—which often mirror traditional peasant practices while bringing current scientific knowledge into their application—suggest that organic methods help reduce agriculture's climate impacts in numerous ways. These include increasing soil's ability to sequester carbon by increasing its organic matter, reducing excess nitrogen from chemical fertilizers, eliminating energy-intensive pesticide production, composting instead of burning crop residues, and feeding ruminants less grain and more grass, among other practices (Muller and Davis 2009).

The need to expand the use of these methods is reinforced by the widespread effects of current and predicted climate changes on growing food. While climate skeptics have argued that crops could benefit from an increase in CO_2 in the air (Gelbspan 1998), scientists have found that increasing CO_2 and rising temperatures combine to increase the rate of plants' nighttime respiration, resulting in a net loss of the metabolic energy acquired from photosynthesis during the previous day. One recent study, examining over 200 farms in South and East Asia, confirmed that rising nighttime temperatures have already systematically reduced rice yields (Welch et al. 2010).

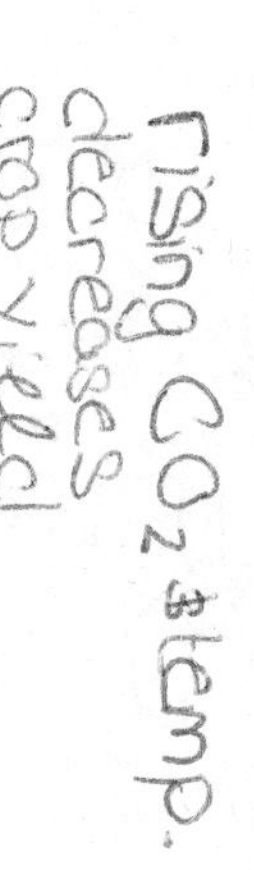

The comprehensive review of current climate science compiled in 2007 by the Intergovernmental Panel on Climate Change (IPCC) addressed several longer-term trends. In the second volume of their report, focusing on the physical, biological, and human impacts of climate changes, the IPCC confirmed that climate-induced disruptions of the global hydrological cycle will bring cycles of increased flooding and droughts, most notably in the major river deltas of Asia and Africa. Glacial melting will continue to affect the water supplies of one-sixth of the global population, that depend on runoff from glaciers to provide much of their fresh water (IPCC 2007, 11).

The data reviewed by the IPCC points toward a worldwide decrease in crop productivity if global temperatures rise more than 3 degrees Celsius, although crop yields from rain-fed agriculture could be reduced by half as soon as 2020. In Africa alone, between 75 million and 250 million people will be exposed to "increased water stress" (IPCC 2007, 13). Agricultural lands in Latin America will be subject to desertification and increasing salt content. Overall, the study confirms that those populations with "high exposure, high sensitivity and/or low adaptive capacity" will bear the greatest burdens, and that those that contribute the least to the problem of global warming will continue to face the most severe consequences (Ibid.). Scientists participating in a symposium published in January 2011, by the British Royal Society, now project "intolerable frequencies of crop failure" in sub-Saharan Africa, potentially as often as every other year in some locations (New et al. 2011, 12). These findings affirm the urgency of embedding a justice-centered framework into global efforts to alleviate the impacts of climate change. Politically, this has become an essential counterpoint to the prevailing, rather narrow, climate policy focus on the reliability of climate models and quantitative projections of future CO_2 concentrations.

Converging Movements

La Via Campesina's increasingly potent interventions into the UN's climate policy process help illuminate the central role of activist farmers in the Global South in the development of a worldwide climate justice movement. How can food and farm activists in the Global North also play a more active role in supporting and strengthening this emerging movement?

As other contributors to this volume have demonstrated, an impressive variety of local and regional efforts throughout the US and other developed countries promote local food, facilitate direct purchases from farmers, and further the goal of community-based food security. Many of these activities, however, take place in the background of a fashionable localism that is often skewed toward affluent consumers. These middle and upper-class individuals often embrace local food as part of a high-consumption "green" lifestyle that at best offers only a pale challenge to destructive patterns of conspicuous consumption.

"Green" consumerism is often driven by corporations that embrace environmental themes in their advertising and public relations, while contributing to the destruction of communities and ecosystems. These include corporations such as Coca Cola, whose president recently told the *Guardian* that the company expected that "as much as 70% of future advertising would have an environmental focus" (Wintour 2009). A northern California farmer quoted in a recent report glibly described a scenario of "islands of good food and good community in a sea of bad news" (Holt-Giménez and Shattuck 2011, 125). A recent *Newsweek* article highlighted the rise of fashionable food activism amid a background of spreading obesity and food insecurity in the US, asserting, "Food is no longer trendy or fashionable. It is fashion" (Miller 2010).

Visionary activists, farmers, and organizers around the world, however, are reaching far beyond the fashions of "green consumerism" and introducing models of solidarity and mutual aid that resonate well with the message of climate justice. For example, neighborhood activists in Hartford, Connecticut, brought an assertive community-organizing model into efforts to alleviate hunger, and developed a comprehensive urban food system. They brought community gardens and farmers' markets into inner-city neighborhoods and developed active working relationships between publicly funded nutrition programs and nearby farms (Winne 2008). Organizations in many other US cities and towns have developed their own innovative approaches to bringing more fresh and locally grown food to those most in need.

New York City now boasts more than 50 farmers' markets, serving neighborhoods at every socioeconomic level, as well as over 30 urban farms and an equal number of public food pantries and soup kitchens that regularly use fresh produce grown in and around the city (Grow NYC; Just Food 2010). In the decaying industrial city of Holyoke, Massachusetts, a large urban farm known as *Nuestras Raices* (Our Roots) offers substantial farm plots to aspiring urban farmers and helps recent immigrants identify varieties of culturally-important crops that are adaptable to the short New England growing season (Gottlieb and Joshi 2010, 123–26). Programs in a variety of cities offer subsidized farm-share subscriptions to low-income residents, often combined with environmental education and job training for neighborhood residents (Henderson and Van Eyn 1999, 204–205). In a more rural setting, a network of nonprofit "food hubs" throughout the state of Vermont is actively linking farmers with local food shelves, hospitals, senior meals programs, and other institutions that once relied exclusively on imported, highly processed foods.

A related challenge has emerged amid the food-cooperative movement in the US. Many consumer co-ops were founded by countercultural activists seeking to bring affordable natural foods into underserved areas, but were dramatically transformed during the 1980s and '90s as natural foods became an important niche in the corporate food system. With the rapid expansion of the organic foods industry and aggressive competition from natural food chains such as Whole Foods, co-op members across the US felt pressured to accept corporate-style management and often saw their stores become pale imitations of their slick, upscale competitors.

A few co-ops around the country survived the competition by becoming more focused on their social mission and community-centered principles, often serving as centers of opposition to genetically modified foods and other corporate excesses (Seydel 2001). They were clearly the exception during that time. But with the rise of local food activism and the recent economic downturn, many co-ops have returned to their original mission of bringing people better food at lower prices, strengthening direct links between farmers and consumers, and involving members more actively in their governance. One initiative, known as the Agricultural Justice Project, links consumer co-ops in the upper Midwest with organic growers and farmworker groups through a Local/Fair Trade label (Henderson and Mandelbaum 2007). Today, with national chains such as Walmart claiming to feature organic and local foods (defining *local* extremely loosely), co-ops in many regions of the country serve as essential bulwarks against the continuing dominance of megacorporations in our food system.

It remains to be seen whether efforts such as these can make a decisive difference in the wider struggles for food sovereignty and climate justice. On a practical level, the jury is still out. Recent research suggests that how our food is grown and processed may have far greater climate impacts than how far it travels (Weber and Matthews 2008). Regionalizing produce, reducing meat consumption (particularly from factory farms), and eating more seasonally can have a greater impact than broadly focusing on the geographic origins of our food. Mega-greenhouses in northern climates that aim to keep tomatoes on our plates all winter are clearly not part of the solution, and curtailing long-distance transportation of staple grains may not be wise either, especially in the short term. However, the development of local food systems that help sustain farmers committed to sustainable agriculture within their regions still ranks among the most immediately practical strategies to enhance both food justice and climate justice. More resilient local and regional food systems contribute to climate mitigation by reducing fossil fuel use, and also help communities cope with future climate instabilities by fending off potential disruptions of long-distance supply chains due to rising energy costs.

Local food systems can also help challenge agribusiness dominance and, ultimately, the climate consequences of agribusiness practices, but only if they move far beyond niche marketing to a selective clientele, toward creating a genuine alternative that serves much larger numbers of people. While more affluent "locavore" activists often shy away from politics, a higher level of political engagement is essential if this movement is to begin realizing its potential. Indeed, many people have come to see local food as part of a broader transition away from dependence on fossil fuels and toward a more fully realized economic self-reliance on both local and regional levels. In a time of climate crisis, as well as mounting food insecurity, can these efforts evolve toward the next step? Can food co-ops, "food hub" organizations, and urban food policy councils help forge working alliances with small farmers, farmworker organizations, immigrant-rights activists, and others, both domestically and internationally, to challenge corporate food monopolies and work toward more fundamental systemic transformations?

Of all the decisions people make every day about their participation in the corporate-driven economy, choices about food are the most personal, and sometimes the most flexible. As we become accustomed to thinking more carefully about the politics of our food choices, we become more fully aware of the implications of our decisions about other key aspects of our home and work lives. But many such choices, for example, how far we may commute to work, are often largely out of our control, and are impacted by much larger decisions, such as how our cities and neighborhoods are designed, and the relative cost of living in various locations. If individual choices can begin to drive different political decisions at the community and regional levels, however, we can begin to challenge entrenched systems of power in society. And if those who are still personally able to make such choices organize to act in solidarity with those who are far less able to do so, we may see the reemergence of a more genuinely transformative social movement.

In his book, *The Green Collar Economy*, Van Jones (2008) recounts an interview with Brahm Ahmadi, executive director of the People's Grocery in Oakland, California, which has also developed community gardens and a two-acre farm. "Food is our medium for achieving broader outcomes in community development and public health and addressing disparities in opportunities and quality of life," Ahmadi explains. He continues:

> We chose food as our tool because it's intimate and universal, regardless of the differences in culture or personal preferences. . . . From there we connect the dots to the structural and systemic issues of the food system: considering the global environmental footprint of food production, how far food travels, and equity issues related to farmworkers and the struggles of small farmers . . . connecting those to the struggles of low-income consumers (Ibid. 130–31).

With the failure of United Nations climate negotiators in Copenhagen and Cancún to reach an international agreement to reduce emissions of carbon dioxide and other greenhouse gases (Tokar 2010; Khor 2010), climate activists are also struggling with the question of what is achievable at the local and regional level. Cities, towns, and entire US states are pioneering important initiatives to reduce energy use and relieve dependence on fossil fuels (Linstroth and Bell 2007), even as advocates for climate justice and food sovereignty are arguing for broader systemic changes.

While many food and agriculture activists understand the need to challenge the power of megacorporations in the global food system, systemic change is perhaps even more central to efforts to forestall climate catastrophe. To curtail the excessive emissions that are threatening to overheat the entire planet requires fundamental changes in almost every area of human activity, including entrenched patterns of energy and land use, the design of our cities and towns, and everyday habits of work and leisure. The survival of threatened ecosystems, and perhaps of complex forms of life on this planet, may now require a pace of technological, political, and economic transformation well beyond anything we have yet experienced.

Ultimately, both food sovereignty and climate justice require greater solidarity and more committed alliance building with people around the world than many in the Global North are accustomed to. Increasingly catastrophic climate disruptions, which now mainly impact people in the tropics and subtropics, are beginning to be felt throughout the world. More than ever, our ability to continue to thrive as humans depends on radically transforming our social and economic systems. The reality is too urgent, and the outlook far too bleak, to settle for anything less. Perhaps more than ever before, we are compelled to realize our vision of a dramatically different kind of world.

Works Cited

Bruno, Kenny, J. Karliner, and C. Brotsky. 1999. "Greenhouse Gangsters vs. Climate Justice." San Francisco: CorpWatch.

Bullard, Robert. 2009. Presentation at "Advancing Climate Justice: Transforming the Economy, Public Health and Our Environment." The 20th Anniversary National Climate Justice Conference. New York, NY, January 29.

Climate Justice Action. 2009. "Only You Can Fix a Broken System: Climate justice movement to converge on UN climate talks." Accessed June 8, 2011. http://climate-connections.org/actions/copenhagen-climate-justice-action.

Climate Justice Action. 2010. "What does Climate Justice mean in Europe? A discussion paper." Accessed June 6, 2011. http://www.climate-justice-action.org/resources/documents/what-does-climate-justice-mean-in-europe/.

Durban Group. 2004. "Climate Justice Now! A call for people's action against climate change." Unpublished.

ETC Group. 2009. "Who Will Feed Us? Questions for the Food and Climate Crises." Ottawa: ETC Group.

Gelbspan, Ross. 1998. *The Heat is On: The Climate Crisis, the Cover-up, the Prescription*. Reading, MA: Perseus Books, 36ff.

Goodland, Robert, and J. Anhang. 2009. "Livestock and Climate Change." *WorldWatch*, November/December 2009: 10–19.

Gottlieb, Robert, and Anupama Joshi. 2010. *Food Justice*. Cambridge: MIT Press.

Grow NYC. 2010. "Our Markets." Accessed November 1, 2010. http://www.grownyc.org/ourmarkets.

Henderson, Elizabeth, and R. Van Eyn. 1999. *Sharing the Harvest: A Guide to Community Supported Agriculture*. White River Junction, VT: Chelsea Green.

Henderson, Elizabeth, and R. Mandelbaum. 2007. "Bringing Fair Trade Home: The Agricultural Justice Project." *The Natural Farmer* (Winter 2007–2008).

Holt-Giménez, Eric, and Annie Shattuck. 2011. "Food crises, food regimes and food movements: Rumblings of reform or tides of transformation?" *Journal of Peasant Studies* 38 (1).

IPCC (Intergovernmental Panel on Climate Change). 2007. "Contribution of Working Group II to the Fourth Assessment Report of the Intergovernmental Panel on Climate Change: Summary for Policymakers." Accessed October 1, 2007. http://ipcc.ch.

Jones, Van. 2008. *The Green Collar Economy: How One Solution Can Fix Our Two Biggest Problems*. New York: HarperCollins.

Just Food. 2010. Accessed November 1, 2010. http://justfood.org/about-us.

Khor, Martin. 2010. "Complex Implications of the Cancun Climate Conference." *Economic & Political Weekly* 45 (52): 10–15.

Linstroth, Tommy, and R. Bell. 2007. *Local Action: The New Paradigm in Climate Change Policy*. Burlington: University of Vermont Press.

Lohmann, Larry. 2006. "Carbon Trading: A Critical Conversation on Climate Change, Privatization and Power." *Development Dialogue no. 48*. Uppsala: Dag Hammerskjold Center.

McIntyre, Beverly D., H. R. Herren, J. Wakhungu, and R. T. Watson, eds. 2009. *Agriculture at a Crossroads: The International Assessment of Agricultural Knowledge, Science and Technology for Development. Synthesis Report: A Synthesis of the Global and Sub-Global IAASTD Reports. Secretariat, 3*. Washington, DC: Island Press.

Miller, Lisa. 2010. "Divided We Eat." *Newsweek*, November 22.

Muller, Adrian, and Joan S. Davis. 2009. *Reducing Global Warming: The Potential of Organic Agriculture.* Emmaus, PA: Rodale Institute.

New, Mark et al. 2011. "Four degrees and beyond: the potential for a global temperature increase of four degrees and its implications." *Philosophical Transactions of the Royal Society A: Mathematical, Physical and Engineering Sciences* 369 (1934).

Ngalame, Elias Ntungwe. 2010. "Immigration surging in Cameroon as farmers and fishermen desert shrinking Lake Chad." Accessed June 8, 2011. http://www.trust.org/alertnet/news/immigration-surging-in-cameroon-as-farmers-and-fishermen-desert-shrinking-lake-chad/

Paustian, Keith et al. 2006. "Agriculture's Role in Greenhouse Gas Mitigation." Arlington, VA: Pew Center on Global Climate Change.

Rosset, Peter. 2010. "Fixing Our Global Food System: Food Sovereignty and Redistributive Land Reform." In *Agriculture and Food in Crisis: Conflict, Resistance, and Renewal.* edited by Fred Magdoff and Brian Tokar. New York: Monthly Review Books.

Saragih, Henry. 2009. "Why We Left Our Farms to Come to Copenhagen." Presentation at Klimaforum 09, Copenhagen, Denmark, December 7.

Seydel, Robin. 2001. "Cooperatives: A Source of Community Strength." In *Redesigning Life? The Worldwide Challenge to Genetic Engineering*. Brian Tokar, ed. London: Zed Books.

Tokar, Brian. 2010. *Toward Climate Justice: Perspectives on the Climate Crisis and Social Change*. Porsgrunn, Norway: Communalism Press.

Weber, Christopher L., and H. S. Matthews. 2008. "Food-Miles and the Relative Climate Impacts of Food Choices in the United States." *Environmental Science and Technology* 42 (10): 3508–13.

Welch, Jarrod R. et al. 2010. "Rice yields in tropical/subtropical Asia exhibit large but opposing sensitivities to minimum and maximum temperatures." *Proceedings of the National Academy of Sciences of the United States of America* 107 (33): 14562–67.

Wightman, Jenifer. 2006. "Production and Mitigation of Greenhouse Gases in Agriculture." Cornell University Agricultural Ecosystems Program Team. Accessed October 25, 2010. http://www.climateandfarming.org/pdfs/FactSheets/IV.1GHGs.pdf.

Winne, Mark. 2008. *Closing the Food Gap: Resetting the Table in the Land of Plenty*. Boston: Beacon Press.

Wintour, Patrick. 2009. "Green consumerism can avert climate disaster, say top firms." *The Guardian*, October 16.

World People's Conference on Climate Change and the Rights of Mother Earth. 2010. "People's Agreement." April 22. Accessed April 26, 2010. http://pwccc.wordpress.com/2010/04/24/peoples-agreement/.

Worth, Robert F. 2010. "Searching for Crumbs in Syria's Breadbasket." *New York Times*, October 13.

Chapter Eighteen

WOMEN'S AUTONOMY AND FOOD SOVEREIGNTY

By Miriam Nobre,
World March of Women

OVER THE LAST 10 years, the World March of Women has adopted a platform for food sovereignty and forged alliances with La Via Campesina and other organizations that advance the concept. We understand food sovereignty to be the right of people, countries, and states to control their own agriculture and food systems. This entails protecting food production and food culture in a way that all persons can access adequate quantities of decent quality food. The challenge for urban women and women of the Global North, is how to participate in this process with the same commitment as rural women and women of the Global South. The path to this goal begins with solidarity among women with different experiences and demands, continues through open debate, and culminates in action with respect to domestic work and caregiving—and action against the commodification of everyday life and of women's bodies.

Our Starting Point

The World March of Women (WMW) is an international feminist movement uniting groups in over 60 countries around a continuous fight to change the world and the lives of women. The movement began in 2000 as a campaign against poverty and sexist violence. At that time, more than five million signatures were collected to support a platform of demands submitted to the

United Nations on October 17, 2000. The first demand was to eliminate poverty by implementing national laws and strategies to ensure women would not be discriminated against in "their rights to access basic resources such as potable water, food production and distribution, in order to ensure food security for the population" (World March of Women 2008, 55).

After this international action, the majority of participating groups decided to continue working together. They established a common agenda, and greater political identity through national coordination, with an international action every five years. The international goals were translated into national contexts, where most countries developed platforms for their own demands.

The second international action was the "Women's Letter to Humanity," which traveled through 53 countries between March 8 and October 17, 2005. As the letter traveled, each participating country provided a cloth square to the Quilt of Solidarity, a visual representation of the letter. The letter is a statement expressing five core values: equality, liberty, solidarity, justice and peace. It reads:

> A society's economy serves the women and men comprising that society. It is based on the production and exchange of socially useful wealth distributed among peoples, with the priority being satisfying collective needs, eliminating poverty and ensuring the balance of collective and individual interests. It ensures food sovereignty.

The letter was drafted over the course of several months before it was finally approved at the Fifth International meeting of the WMW in Kigali, Rwanda. In 2005, the replacement of the term *food security*, used in 2000, with *food sovereignty* probably occurred due to open participation in the drafting process. The fact that rural women appropriated the principle of food sovereignty when La Via Campesina proposed its emergence in the "Letter to Humanity" is an indication of the influence of grassroots organizations on the overall movement.

As the letter and quilt traveled from country to country, small rural towns proved to be important stopping points and protest zones. Local groups from these areas were responsible for organizing activities around the letter and generating national recognition from other leaders. This shattered the traditional image of the feminist movement being led by urban women living in large urban centers. The new dynamic was reflected in the letter's demands, which emphasized rural women's concerns.

At the end of the letter's journey, we consolidated different countries' demands around common goals, identifying four fields of action: women's work, the common good, violence against women, and peace and demilitarization. These guided the WMW's international agenda from 2006 until 2010, when we reexamined them. At first, the term *common good* referred to the fight against the privatization of natural resources and to women's community rights to define the use of land, water, biodiversity, and food sovereignty. In 2008, this term was expanded to include education, health, common community knowledge, and the fight against its privatization.

all rights!

Learning to Work in Alliances

In 2007, in conjunction with La Via Campesina, Friends of the Earth International, and other organizations, we organized Nyéléni, the Forum for Food Sovereignty, in Selingué, Mali.

In one of the first workshops held by the WMW and La Via Campesina, one of the women farmers challenged the group, saying, "The issue is that we have a core difference. We want to preserve our space and time in the kitchen and in the preparation of food, which is an expression of our culture and our knowledge, and prevent it from being replaced with junk food, and you want to stay far away from the kitchen."

From the beginning, we realized we operated on contradictory ground; we value caregiving, which is made invisible by the patriarchal and capitalist system, yet we do not want to do it alone. We want to share this work with men and, collectively, with social organizations and the state, with public policies that support caregiving. We want to stop society from overworking women so we have time to pursue our own interests, yet we also reject market solutions like fast, industrialized food.

We believed Nyéléni should focus on helping women from different sectors (farmers, fisherwomen, pastoralists, migrants, etc.) affirm themselves as political entities with their own analyses and demands. The most prevalent theme in our debates was women's right to land, water, seeds, and territory, and their role in the production, preparation, and distribution of food. The Nyéléni women's declaration agreed to "reject capitalist and patriarchal institutions that conceive of food, water, land, people's knowledge, and women's bodies as mere goods."

In 2008, the VII International Meeting of the WMW in Galícia, Spain, produced a mass demonstration in front of an international supermarket chain, an open forum for debate in a public market in the center of Vigo. Our Galícian sisters evaluated the collective work of the WMW, La Via Campesina, Friends of the Earth, and other ecological and consumer movements positively. Our achievements resulted in a demand that the concept of food sovereignty be incorporated into the Constitution of Autonomy of Galícia. The WMW encouraged participation in a local, responsible consuming cooperative. In 2010 a WMW Galícian organization was at the forefront in a countersummit questioning European Union fishing policies, drawing attention to the fierce exploitation of women working in the fishing industry.

Food sovereignty is a popular concept in Latin America, Africa, and Asia but is less prevalent in North America and Europe. Globally, the concept still mobilizes more rural than urban women. To forge new alliances between rural and urban women, we must deal with the following issues: caregiving in society, referred to here as reproductive work, and the commodification of everyday life in relation to women's bodies, as defined by a system of patriarchy.

Politicizing reproductive work

Food sovereignty constructs a political agenda around reproduction that involves everyone, not only women. Reproduction, caregiving (of children, sick persons, seniors, and men), and assuring food, health, and general welfare are considered women's tasks. Even more than tasks, these are considered the core of a woman's identity. Being a woman means that one must always be ready to attend to others' physical and emotional needs—needs we all have throughout our lives.

The market production of goods and services is the only sphere in which policy debates and social movements tend to direct their economic demands. However, the production of goods would be impossible without work, and for this we must be fed, cared for, and materially and socially reproduced. In order to bring to light that which is hidden, we feminists talk about a sexual division of labor and two economic spheres, production *and* reproduction.

In general, the production sphere is considered masculine and the reproduction sphere is considered feminine. But these spheres are not separate, and we seek to reveal the links hidden between them. For example, when Structural Adjustment Programs demand governments cut social programs, the social work itself does not disappear, but is shifted to the sector of nonwage work done by women in their families and communities. When companies function in the "just in time" logic, they hire women in times of harvest or for bulk orders, but then fire them when this period ends. Women's unemployment is not a problem, since 'women always have so much to do in the home, anyway.'

It is essential to understand the sphere of reproduction in its own logic, not as an inverted mirror of the sphere of production. For example, simply counting the time dedicated to caregiving does not reveal its full dimensions. To begin with, many activities are simultaneous, so the question is instead, how women manage and prioritize the work (for example, watch the stove, help with the children's schoolwork). Further, more than the question of getting work done is the permanent disposition to get it done, from guessing the wants and needs of children and spouse, to predicting whether or not there will be enough sunshine during the day to wash and dry clothes.

The feminist economist Cristina Carrasco (2008) has analyzed the time and logic of caregiving, concluding it is irreconcilable with the time and logic of the market.When the two interact, it is at a great cost to women. Women's work is the adjustable variable that maintains the margins of exploitation of labor in order to accumulate capital, i.e., profit. Contrary to the proposal of companies, states, and international institutions, we are not in search of a conciliatory policy between paid work and the family, but rather seek to overcome the logic of the market. This framework, largely contributed to by feminist economic theory, allows activists from women's movements to create and utilize increasingly complex analyses in moments of change. It allows us to ask: "How do we build rights and claims in terms of food sovereignty? And how do we succeed in winning these rights?"

Just as we have a long history of demanding services that support reproduction, such as daycare centers, school food, and public Laundromats, we also have substantial experience in assuming collective processes in reproductive work. Women organizing into shopping groups or engaging in collective food preparation are common in the working history of the movement, even if this is not very visible. Throughout every long strike or massive layoff, one finds a group of women keeping working families fed.

In the political economy of resistance, there is always food collectively prepared by women, such as the *ollas comunes* (communal pots) in Honduran communities, that are declared a coup-free zone. It is these same women who participate in the *ollas comunes* that in protests add the motto *Ni golpes de estado, ni golpes contra las mujeres* (No coup d'etats or violence against women).

In Latin America, there are many accounts of women getting together in communal kitchens, buying groups, and milk-distribution groups. The majority of these groups are created in response to moments of crisis or extreme levels of poverty. With some exceptions, such as in Peru, these experiences are dispersed, and not considered by the feminist movement to be part of the collective history of the women's movement. This is because the nucleus of these organizations is a traditionally feminine activity and therefore considered a process which supports women's subordinate role. However, in the WMW, we believe that there is a big difference between being responsible for food preparation in individual homes, and the collective preparation of food. Women often begin organizing in precarious conditions in order to also attend to their responsibilities as mothers and later, as a result, break out of their traditional, subordinate roles. They occupy public space, negotiate with authorities, and question the established order in their community and in their families.

Therefore, our challenge is how to make social movements understand the importance of social reproductive politics and act upon them. In other words, how do we raise awareness and change personal and collective practices *within* social movements? We must also determine how to lobby our governments to implement structural changes, such as the way in which cities, public transportation, and official workdays are organized.

We are women, not commodities!

When we began meeting with rural women fighting against genetically modified organisms (GMOs) and pesticides, we saw that the international companies that were advertising and lobbying to distribute GMOs were the same companies that produce synthetic hormones that promise eternal youth, or contraceptives that are beyond the control of women, such as injections or hormonal implants. We realized that the so-called "bioindustry" and the associations of industrial manufacturers of agricultural inputs, seeds, food processing, and drug manufacturing, have increasingly similar strategies. The use of nanotechnology in agriculture and cosmetics is yet one more example of the strategies used by these companies.

From our perspective, we should take into account how these strategies have developed and how they are organized in our daily lives, and consider how to find alternatives. At the center of these alternatives is a new understanding in the relationship between reproductive work and overcoming the alienation between our bodies. Making peace with our bodies, which are represented in

Without Women There Is No Food Sovereignty

By Esther Vivas

In diametric opposition to capitalist neoliberal policies' negative impact on peasants, and peasant women in particular, La Via Campesina's alternative proposal to the agroindustrial food system seeks to incorporate the voices and concerns of women. Though women are the primary producers of food worldwide, they often occupy an invisible space in the patriarchal corporate food regime, as their roles are considered subordinate to men's, and access to land tenure remains equally limited to women in both the Global North and South. Despite the "femininization" of agroexport industries, women's salaries as agricultural workers are significantly less than men's, yet women take on the dual role of working and maintaining a family. The neoliberal doctrines evident in structural adjustment policies of the 1980s and '90s had an especially deleterious effect on women's access to health care, housing, and education. These policies further exacerbated longstanding patriarchal modes of physical, social, and economic violence against women.

The 2008 price spikes in basic foodstuffs put in evidence the vulnerability and unsustainability of the corporate agroindustrial food system. Food sovereignty movements provide an alternative to a top-down, corporate food system and reclaim the right to choice—choice in what, how, and where we produce our food—effectively putting power back into the hands of peasant women and men. In this context, La Via Campesina incorporates the feminist voice in addition to promoting networks of solidarity among women worldwide. Women occupy dual roles within the food sovereignty movement as both dynamic organizers and participants, reaffirming their commitment to transform not only the corporate food regime but deeper patriarchal structures of violence. Thus, it is on these grounds that food sovereignty comprises not only a break with our current capitalist model of agribusiness, but also a break with the patriarchal system that oppresses and dominates women.

Full Article at: http://www.foodmovementsunite.com/addenda/vivas.html

the patriarchal system as fragile, sickly, unstable, is essential in feminism.

The exploitation of our work and our time in order to generate profits for the few creates suffering. We attempt to reduce this suffering with quick remedies. Medicines that regulate demeanor, such as antidepressants, ensure profits for the pharmaceutical industry. Our relationship with food is similar; we feed our anxiety with sugars and carbohydrates. Genetically engineered organisms, food additives, and vitamin supplements all turn basic nutrition into a collection of quick-fix remedies. We depend on doctors' and specialists' recommendations for our health. However, medicine, like all other sciences, operates within androcentric patterns. Women's bodies are only specifically considered in terms of pregnancy and birth. For example, the fact that women are more sensitive to agrotoxic contaminations due to the characteristics of their bodies, is ignored (Boston Women's Health Book Collective 2000).

In terms of eating disorders women are overrepresented in the population. This is not just a public health concern but a theme for political debate about how our society relates to food, and the requirements and controls that take place upon and in women's bodies (Arnayz and Comelles 2007). The imposition of a standard of beauty and the ideology of value as attributed by the gaze of others, especially men, makes women vulnerable to the cosmetics industry and plastic surgery (the most evident commodification of women's bodies). The "perfect" body can now be purchased, in order to find or keep a companion, sell oneself in the industry of prostitution, or even to get a company job requiring "good appearance." What these motivations have in common is their distance from the personal right that is expressed in the motto *Meu corpo me pertenece* (My body belongs to me). Contrarily, they all respond to the imposed constraints and expectations on women in relation to a patriarchal society.

Women are questioning the relationships between commodification and their bodies, between themselves and others, and between women and nature. The homogenization of standards of feminine beauty is similar to the homogenization of crops found in industrial monocultures. We therefore search for other paradigms with which to organize our daily life and the production and reproduction of our society.

The Current Construction of the WMW Regarding Food Sovereignty

Food sovereignty has been integral to WMW activity in several countries. In 2007, a WMW campaigning for rural women's rights in India sought to strengthen their involvement in food production and agricultural work. In Turkey, WMW groups became involved in a campaign against the water privatization efforts spearheaded by the Coca-Cola Company. They are part of the national platform against GMOs, framing their food sovereignty goals from a feminist perspective. In Mali and Benin, women's associations act collectively in market choices. Their aspiration is to develop markets in West African countries that exist as an alternative to the free-market rationale. In Peru, women participated in a campaign for native seeds and defied water-privatization efforts.

The WMW Third International Action of 2010 recounted these experiences and opened up an international debate. In Belgium, a demonstration of over 6,000 women demanded, among other things, that women in agriculture be recognized as having social rights, whether they are farmers' wives or migrant laborers. In addition, they protested the North's economic policies, highlighting their consequences on the whole planet and especially on the Global South.

We wish to go beyond debate. In good feminist tradition, we believe collective experience breeds strong movements. We stress the importance of food in political organization. Food preparation is a class issue as well as a gender issue. Many middle-class women do not concern themselves with the labor required to feed everyone at public events. We all run the risk of replicating society's division between "professional" work (methodology and discourse) and manual labor (logistics and food). Our movement is assuming more and more collective responsibility for food preparation and distribution.

It is already obvious to food movements that they should prepare their own food, using agroecological products bought directly from small producers. However, the same cannot be said of movements led by women, who cook every day for their families or for the families for whom they work. The feminist movement provides them an escape from these responsibilities and time for themselves. The question is how can we maintain this respite while collectively assuming food-preparation responsibilities?

While 2,000 Brazilian sisters marched between March 8 and 18, 2010, 80 women marched in the kitchen, preparing meals and holding debates. Every day, a team of 20 women took turns washing up and participating in debates. The Brazilians arranged to spend one day of the march in the kitchen, to learn how to run it while also training in political debate. This proved challenging; not only was there limited experience among participants but limited resources as well. Still, our sisters acted respectfully, staying true to popular feminist education principles. They valued the group's diverse backgrounds, ages, and experience, as well as the absence of a chef. When the march arrived at the kitchen site, we were received by our Brazilian sisters with the slogan "*A cozinha é o coração, sem comida não há revolução*" (The kitchen is the heart, without food there is no revolution).

Overcoming Obstacles to Food Sovereignty

We united behind the principle of food sovereignty first because our rural sisters in the World March of Women invited us to join their struggle for land and fair conditions to live and produce as farmers. Secondly, as allies of La Via Campesina and Friends of the Earth, we understood the importance of uniting all groups dedicated to improving living conditions for both men and women. We also understand that food sovereignty allows us to expand the feminist movement's horizons. Furthermore, questioning the system of modern consumption from the starting point of what we eat brings us closer to our own bodies, which are alienated, mistreated, and reduced to mere commodities.

Food sovereignty opens doors to other issues. It urges us to address energy sovereignty, and sovereignty over the territory where we live. The concept of territory encompasses the right to land, water, biodiversity, and self-determination. We believe our bodies are our primary territory. To live in pleasure and harmony with one's body, free from the threat of physical violence or harmful consumption, is a political act. As we fight to defend our territory from GMOs, we fight to rid it of violence against women.

The strength of the food sovereignty movement comes from its linkages with other movements. Our contribution as a feminist movement is to link the goal of women's autonomy with the vision of sovereignty for all people.

Our Common Agenda: Demands and Commitments of the World March of Women Third International Action

In the struggle for common goods and access to public services we demand:

- The promotion of alternative, clean energy sources (biodigesters, solar and wind energy) and the rejection of nuclear energy, as well as the democratization, decentralization, and public management of energy in ways that will guarantee the rights of all peoples, including those of indigenous peoples;
- Universal access to drinkable water and basic sanitation, as well as public services of quality (health, education, public transport, etc.), provided by the state acting as guarantor of basic rights;
- Agrarian reform and the promotion of agroecology (organic agriculture, etc.), in opposition to the privatization of the

environment—and the abolishment of all barriers preventing rural communities from saving, preserving, and exchanging seeds among themselves, their countries, and continents;

- Severe penalties for industrial countries and transnational companies responsible for the contamination and destruction of our environment and changes in the food chain, as well as immediate measures to stop this situation;
- Reparation of the ecological debt owed by industrialized countries, most of which are in the North, to peoples in the South. This debt has been incurred via the gradual appropriation and looting of natural resources, and the abusive appropriation of communal spaces such as the atmosphere and the oceans, which has created numerous socio-environmental problems at local levels;
- Support for countries where the consequences of climate change and intensive, chemical-based agriculture have amplified the effects of natural disasters.

And we commit ourselves to the following:

- To affirm the principles of, and strengthen the struggle for, food sovereignty;
- To deepen our analysis of the access to and consumption of energy;
- To establish and strengthen links among urban and rural women through direct-purchase experiences, fairs and collective food preparation and distribution; to exchange knowledge and ensure that the "urban point of view" is not privileged with regard to analysis and practice; to struggle for a change in eating habits, shifting from imported junk food to locally produced, healthy food; and to denounce the hegemony of the agribusiness industry and the big supermarket chains in food distribution;

- To identify and denounce transnational companies that undermine food and energy sovereignty;
- To denounce market solutions to climate change, such as the clean-development mechanism, joint implementation, and emissions trading schemes (the three main pillars of the Kyoto agreement);
- To hold peoples of the North accountable for their consumption and lifestyle, and struggle for changes in consumption and production models relating to goods, food, and energy; and to raise awareness of the need to reduce Northern demand for resources from the South.

Works Cited

Arnayz, Mabel, and J. Comelles, eds. 2007. *No comerás. Narrativas sobre comida, cuerpo y género en el nuevo milenio.* Barcelona: Icaria.

Boston Women's Health Book Collective. 2000. *Our Bodies—Ourselves*. Simon & Schuster. New York

Carrasco, Cristina. 2008. "Por uma economia não androcêntrica: debates e propostas a partir da economia feminista." In *Trabalho doméstico e de cuidados. Por outro paradigma de sustentabilidade da vida humana*, edited by Maria Lúcia Silveira and Neuza Tito. São Paulo: SOF.

World March of Women. 2008. "Demands of the World March of Women in the Year 2000 in The World March of Women 1998-2008 A Decade of International Feminist Struggle." World March of Women. Sao Paulo, 2008, 55

Chapter Nineteen

TRANSFORMING OUR FOOD SYSTEM BY TRANSFORMING OUR MOVEMENT

From a Conversation with Rosalinda Guillén,
Community to Community

IF WE CAN'T RESPECT ourselves enough to feed ourselves food that is not damaged and, most of all, is not hurting something else in the production of that food, then what are we? Are we really human, then? We know we are doing something wrong; how do we get millions of people to understand that?

I firmly believe that this whole idea of how we build movements in the United States has gone terribly awry; we think that we are doing the right thing and we're really not. I know that we're not because I know by the suffering of the farmworkers in this country; I know by the suffering of the poor, and the lack of food in many areas. What's worse is that through being fed, we're becoming ill.

Our culture is being destroyed by the food we are being given to eat, and our culture is being destroyed when Mexican farmworkers are stopped from feeding themselves the way they want to be fed, which, many times in rural or in urban areas, requires that we grow our own food because we can't find what we need to eat—that we *need* for our own self-sustenance and that is not permitted in many areas. So lately, through working with the US Alliance for Food Sovereignty; I've been really thinking through what food sovereignty is ... that food security, *really*, and food justice are just not good enough for us in this country, and that we really do need a movement toward food sovereignty. But again the question keeps being asked: "What does that mean, a movement toward food sovereignty?"

There are health issues from the lack of food in some areas, the health issues from too much of the wrong food, and the situation of the food-production workers are just not being recognized as elements in the food system. Workers have been saying for many years that there is something wrong in the food system. We're getting sick. We're being forced to do things that are not right. Similar to what the oil workers on the Gulf Coast have been saying: they've been forced to do things that they believe were not right. They've been sounding the alarm, and we haven't been listening. It's the same with food workers all across the food chain.

To me, food justice is about how we as humans are going to take responsibility for feeding ourselves in a way that doesn't hurt people, in a way that doesn't hurt the land, that doesn't hurt the resources that we need so badly just to survive. The need to make a change quickly is accelerating, and I think it's because we've ignored so much for so long. It's not because we didn't know about it. Cesar Chavez was telling us about it in the 1950s and '60s—millions of people stopped eating grapes for almost a whole generation because people understood when he said that there's something wrong with the way these grapes are being grown, and the best way for you to help us correct that is to stop eating the grapes. I think that food justice has so much to do with ourselves as human beings, period. I think this is taking longer than anybody ever hoped, what Cesar hoped, what Martin Luther King hoped. It's taken longer. It shouldn't. But how are we going to take responsibility? *Everything we put into our mouth makes a difference.*

We don't seem to have the sense or willingness to sacrifice to make the difference. It's difficult to pull people out of their comfort zones. We used to pull people out of their physical comfort zones by showing them how farmworkers were living in the labor camps like organizers did in the '60s, but now it's the emotional comfort zone: "Don't call me racist, don't call me privileged, don't call me insensitive, because then you're moving into my comfort zone of who I believe I am as a good person." Even that is becoming a problem; to really tell people the truth we have to speak truth to power. The power of the corporation to influence the emotional being of people is becoming a barrier to us as food justice activists.

So in Community to Community, we are trying to learn from the social forum model and say, based on that model of creating space and dialogue and intersecting movements, can we in the United States develop a women-led organization that replicates that model in a smaller way in local communities? In our case, we are trying to do it by intersecting regionally and then linking nationally. Even our own internal structure and organizing dynamics are an experiment, as well as everything else that we are trying to do to move farmworker justice forward. Women's leadership is our first goal; farmworker justice is next; and then immigrant rights, environmental justice. We are looking at it through the ecofeminist lens.

One of the things we talk about in terms of ecofeminism is building power—and I don't even know if *power* is the right word. A lot of organizations and a lot of organizers talk about power, how we're going to "get the power." They talk about shifting power. It's like, somebody's got power, I want it, our people need the power that that person has or that group has. I don't think our organizing work should be about that. We don't want to take somebody's power. We don't want to be in that place of power. We want to transform what power is. What *is* power? We need to claim that and act on it, regardless of who is claiming it in that other way.

So, a transformation of what power is should be the ultimate goal of community organizing today. Simply shifting power from here to this person, or to this group, doesn't change the structures. It doesn't change the systems. It just places our people in losing situations. I think that we've done that over and over again. We say we were for that person and that person moved along and was elected and then sold out. But they haven't sold out. You have just placed them in a box for them to bang their head against. They're never going to get to you because you moved them out of here, and put them in there, and now you're all alone. It's just so clear to me, but I talk to other organizations and other leaders and try to talk about how we organize, and our model of organizing, and they say, "Well you know that's a wonderful dream, but we've got to get to this, we've got to change this regulation, we've got to change this law," and I'm thinking, *we're inside that box*.

We have to find a way of using our current system to move ourselves out of that structure and create a new one, or just forget about it totally. I think those are the decisions we need to make as organizers and leaders now. So that's what I'm trying to teach young people: Don't fall into that trap! I'm challenging them, I'm speaking to students as much as possible and saying: if out of 300 of you I can get one to move out of this box and go over here and start thinking differently, then today is a successful day. I think it's time we did things differently.

Community organizers really need to be effective at modeling, and being public in modeling what we believe another world should be. We have to model those behaviors, the way we believe the earth should be and other humans should interact, which means a lot of very public presentations and public demonstrations of how we should behave as humans. It's been done in the past. All experienced organizers will tell you, you just need a really strong core moving things quickly, with a lot of people around you. Each and every single movement that you see that has brought change and transformation—more than just legislative change but transformation of relationships between humans in a community—has included music, food, orators, a lot of dialogue, a lot of activity.

How do we convince all the people around us that know nothing about what's in their food, to think about what they're putting in their mouths? I think we can use the tools of the Internet and all the other tools, but I think that our first major sacrifice is figuring out: How do we have that face-to-face contact to really feel the full strength of each other's thoughts and commitments and feelings and goals? Why are we doing what we're doing, and why are we doing it the way that we are doing it? Because the only way it's going to work is for us to be able to understand each other. And it's going to take a long time. By a long time, I mean concerted effort over years—if we can make that effort. But we have to be honest with each other. I think that's a first step: if we can really say what we think in a way that is respected. There is oppression in this country that is very real but you don't know how to articulate it because in many ways talking about oppression is not permitted, but you are not told it is not permitted. It is just not permitted, so you live your life and you don't succeed the way that you want, but you think it's your fault or it's something that you've done wrong.

I think that food sovereignty is the best way to describe our struggle because it speaks more to the dignity of the human person. While food justice is a great term and a great struggle, it speaks more to a struggle based on legislation, policy regulation. It's become a way of struggle that needs to be fought within the existing structures that we recognize. Who ensures justice, if not the same government and corporate food system that is depriving us of our human right to healthy food that does not harm people or the environment?

The political struggle in this country today is controlled. It's almost like there is a monopoly on it. I see ourselves banging our heads on walls that just don't give way. I envision this soft metal box that we're in. We think that reform is putting a dent in the structure we're being held in. We're banging our heads and scratching them a little bit and thinking, "Victory, victory!" I think there is a way out of that box but we don't see it because we believe in the box. It's going to be hard to break through that consciousness that's been created through several generations in this country now. So, new laws and the enforcement of existing laws is important, but it is not enough if we want to transform our food systems.

Food sovereignty demands that we move out of that box and think as human beings of our own personal dignity, and the dignity of our communities, in a deeper, transformative way. What is it that I need to do to ensure my community's liberation, not just from the effects of oppression—like bad treatment of workers and food insecurity—but from *the structures of oppression.* Some of the older activists in this country, like Grace Lee Boggs, understand this. We need to get people to listen to her to grasp what this struggle means before it's too late.

The damage that's done to the earth is getting worse, but we're not stopping it. We are putting poison in our mouths that the corporations are giving us, and we're not saying anything. We're just eating more of it, because it's cheaper or convenient. I think that if we can think of every piece of food that we put in our mouths, that it has to be honored by the fact that we know nothing's been hurt or damaged by that food, in so doing, it will be good for us and it will be good for the earth. People say, "That's impossible! What kind of campaign is that?" It's *not* a campaign; it's more than that. Stop! You're back in your box! We're working within the box that has been given to us by the corporations. They already have that system down; it's theirs. We have to do something else.

There are rules and restrictions. Not just legislative rules, either, but social and cultural conformity about how we're supposed to be. In the social justice movement, we have organizing models and organizing protocols, and you do this and you don't do that. By breaking out of that we really take the ultimate action that's going to move us out of the box.

I think that we really need to rethink all of our relationships when it comes to the food movement. Some folks from Brazil have an interesting thought about coalitions, networks, about movements, about relationships and how we relate to each other to create that political will to make change. Manuel De Landa (2006) has introduced a useful distinction between two general network types:

hierarchies and flexible, nonhierarchical, decentralized, and self-organizing "meshworks." This is an articulation of something that for many years we, and the farmworker community, starting with the farmworker movement that Cesar Chavez started in the Central Valley of California, did. We self-organized through chaotic networks, these chaotic verbal communications and face-to-face meetings of farmworkers, where emotions come into action from the bottom up and create masses of people coming together. We saw it again in the immigrant rights movement a short time ago, with the marches that happened. This kind of meshing of people coming together in this country gets swallowed up and created into hierarchical "networks." And then whatever it was we started from the beginning dissolves into something totally different that all of a sudden needs to be "funded," needs to be "organized" and needs to be "directed."

I think that's part of the problem that's being created in our country, where grassroots movements coming from the bottom up, made up of folks who with their hearts and their spirits, seek to change something wrong in American society and are swallowed up by these other structures. We immediately get put into a square. We must think through what that means and how we might be able to get out of these "containments" of organized movements and really free that up into some sort of thinking of how we create these "meshworks." Even better, we must recognize when this grassroots meshwork is coming up and, for God's sake, leave it alone! Let it grow! Adopt a position of solidarity that in this country has created so many successful social change movements.

If anything requires great sacrifice from all of us, it is the food movement, because every day each one of us puts into our mouths something that has hurt somebody, has poisoned the earth and continues to create that kind of damage to Mother Earth, and we are eating it. At some point we are going to have to say, "No more. We have to stop that. We have to stop eating food that is hurting another person or is hurting the earth." And that, to me, is going to be the greatest sacrifice that we all can make, and when will we get to that point? Hopefully it won't be too late.

Works Cited

De Landa, Manuel. 2006. *Real Virtuality Meshworks and Hierarchies in the Digital Domain*. Netherlands: Netherlands Architecture Institute.

Synopsis

FOOD MOVEMENTS UNITE!
MAKING A NEW FOOD SYSTEM POSSIBLE

By Eric Holt-Giménez and Annie Shattuck

A CRISIS, wrote Antonio Gramsci, happens when the old refuses to die and the new cannot be born. The corporate food regime may not be dying, but it is cracking, as new food systems struggle to be born.

The food riots of 2008 that swept through the Global South returned with runaway food-price inflation in 2010–2011, this time sparking full-scale rebellions in Tunisia, Yemen, and Egypt. Unable to control price inflation or contain rebellion, the oligopolies of the corporate food regime are trapped in a classic crisis of capital accumulation. Monsanto Company—voted company of the year by *Forbes* magazine in 2008—has saturated its Northern markets. Its new gene-stacked products are performing poorly and the expiration of its patent on Roundup has opened the door to Chinese competition. In the face of precipitously falling profits and stock values, the seed giant—along with 16 other monopolies—is trying to use the food crisis as a lever to break open markets of the Global South. This help comes in the form of public-private partnerships of government aid campaigns like the US's "Feed the Future" initiative, and projects to prepare food-deficit countries for the spread of GMOs, like the Bill and Melinda Gates Foundation's Alliance for a Green Revolution in Africa.

The global recession has exacerbated the desperate conditions of the so-named "Bottom of the Pyramid" (BOP): the world's poor living on less than US $2/day—70% of whom are peasant farmers. With the food crisis as their rationale, the world's agrifoods monopolies are jockeying to capture the BOP market. Even though the poor do not spend much individually, they number over 2.5 billion and as a market sector are growing at the rate of 8% a year. In the Global North, retail giants like Kroger, Walmart, and Tesco are scrambling over each other to acquire cheap urban land in the inner cities of the US. Having saturated the rural and suburban markets, these corporations are expanding their operations through tax breaks, government stimulus monies, and political support from First Lady Michelle Obama's campaign to "eradicate food deserts." But this terminology is deceiving. In fact, the food dollars from these areas are significant. In West Oakland, California, 50,000 low-income residents spend over half a million dollars a year on food—dollars that if recycled through locally owned retail could contribute significantly to community economic development. The term *food desert*, like the term *unused land* in the Global South, is used to justify the corporate expansion into land and economies where people make their livelihoods. Like the infamous Wall Street bailouts of 2008, the "solutions" to the global food crisis are actually designed to solve the financial problems of the world's oligopolies.

But the global food crisis is more than the tragic increase in the number of hungry people and the pandemic of diet-related diseases. It is more than the violence of land and resource grabs, the loss of rural livelihoods, and the abuse of food workers. It is more than the cyclical crises of capital accumulation experienced by the world's agrifood monopolies. The food crisis is a *political* crisis.

For this reason, ending the crisis requires more than simply producing more food or making healthier choices. Ending the food crisis is a political project requiring social, economic, and political organization for transformative change. Many food movement organizations are well aware of this, others increasingly so. How can we turn the food movement into a political force for transformative change, rather than just a passing fad, a basket of weak reforms, or a collection of isolated food and agriculture projects?

The farm, food, and labor activists contributing to this volume have addressed this question by calling for political convergence. In doing so, they remind us that it is not enough to have good ideas, good practices, or even good analyses. Forging healthy, equitable food systems requires more than adding to the growing mix of innovative agroecological practices and localizing food or good food policies. Food sovereignty, food justice, and the right to food ultimately all depend on building a unified food movement diverse enough to address all aspects of the food system, and powerful enough to challenge the main obstacle to food security—the corporate food regime.

The Corporate Food Regime

A food regime is a rule-governed structure of production and consumption of food on a world scale. The first global food regime spanned from the late 1800s through the Great Depression and linked food imports from Southern and American colonies to European industrial expansion. The second food regime reversed the flow of food from the Northern to the Southern hemisphere to fuel Cold War industrialization in the Third World.

Today's corporate food regime is characterized by the monopoly market power of agrifood corporations, globalized meat production, giant retail, and growing links between food and fuel. This regime is controlled by a far-flung agrifood industrial complex made up of huge monopolies including Monsanto, ADM, Cargill, and Walmart. Together, these corporations dominate the governments and the multilateral organizations that make and enforce the regime's rules for trade, labor, property, and technology. This political-economic partnership is supported by institutions like the World Bank, the International Monetary

Fund (IMF), the UN World Food Program, USAID, the USDA, and big philanthropy.

Liberalization and Reform

Like the capitalist economic system of which they are a part, global food regimes historically alternate between periods of economic *liberalization*, characterized by unregulated markets, privatization, and massive concentrations of wealth, accompanied by devastating economic and financial busts—the costs of which are paid for by citizens, consumers, workers, and taxpayers. This eventually leads to social unrest, which, if widespread, threatens profits and governability. Governments then usher in *reformist* periods in which markets, supply, and consumption are reregulated to reign in the crisis and restore stability to the regime. In cases where governments are incapable of reform—as witnessed in 2011 in Egypt and other countries in northern Africa—rebellion and revolution can result.

Infinitely unregulated markets would eventually destroy both society and the natural resources that the regime depends on for profits. Therefore, while the "mission" of reform is to mitigate the social and environmental externalities of the corporate food regime, its "job" is identical to that of the liberal trend: preserving the corporate food regime. Though liberalization and reform may appear politically distinct, they are actually two phases of the same system. While both tendencies exist simultaneously, they are rarely, or only briefly, ever in equilibrium, with either liberalization or reform hegemonic at any period of time.

Reformists dominated the global food regime from the Great Depression of the 1930s until Ronald Reagan and Margaret Thatcher ushered in our current era of neoliberal "globalization" in the 1980s. This phase has been characterized by deregulation, privatization, and the growth and consolidation of corporate monopoly power in food systems around the globe.

With the global food crises of 2007, 2010, and 2011, desperate calls for reform have sprung up worldwide. However, few have been forthcoming, and most government and multilateral solutions simply call for more of the same policies that brought about the crisis to begin with: extending liberal (free) markets, privatizing common resources (like forests and the atmosphere), implementing technological "fixes" like genetically modified seeds, and protecting monopoly concentration. Collateral damage to community food systems is mitigated by weak safety nets—be that food aid from the World Food Program or food stamps from the USDA. Unless there is strong pressure from civil society, reformists will not likely affect (much less reverse) the present neoliberal direction of the corporate food regime.

Food Enterprise, Food Security, Food Justice, Food Sovereignty

Combating the steady increase in global hunger and environmental degradation has prompted government, industry, and civil society to pursue a wide array of initiatives framed within discourses of food enterprise, food security, food justice, and food sovereignty. Some efforts are highly institutionalized, others are community based, while still others build broad-based movements aimed at transforming our global food system. Understanding which strategies work to stabilize the corporate food regime and which can actually change it is a fundamental task for political convergence.

As evidenced in this volume, some actors within the global food movement have a *radical* critique of the corporate food regime, calling for food sovereignty and structural, redistributive reforms including land, water, and markets. Others advance a *progressive* food justice agenda, calling for access to healthy food by marginalized groups defined by race, gender, and economic status. Family-farm, sustainable-agriculture advocates and those seeking quality and authenticity in the food system also fall in this progressive camp. While progressives focus more on localizing production and improving access to good, healthy food, radicals direct their energy at changing regime structures and creating politically enabling conditions for more equitable

POLITICAL TRENDS		
CORPORATE FOOD REGIME		
POLITICS	**Neoliberal**	**Reformist**
DISCOURSE	**Food Enterprise**	**Food Security**
Main Institutions:	International Finance Corporation (World Bank); IMF, WTO: USDA (Vilsak); USAID; Green Revolution Millennium Challenge; Global Harvest; Bill and Melinda Gates Foundation; Cargill, Monsanto, ADM, Tyson, Carrefour, Tesco, Wal-Mart	International Bank for Reconstruction and Development (World Bank); USDA (Meerigan); USAID (Feed the Future) FAO; CGIAR; mainstream Fair Trade; many philanthropic foundations and development programs. most food banks and food aid programs.
ORIENTATION	Corporate/Global market	Development/Aid
MODEL	Overproduction; Corporate concentration; Unregulated markets and monopolies; Monocultures (including organic); GMOs; Agrofuels; mass global consumption of industrial food; phasing out of peasant & family agriculture and local retail.	Mainstreaming/ certification of niche markets (e.g. organic, fair, local, sustainable); maintaining northern agricultural subsidies; "sustainable" roundtables for agrofuels, soy, forest product, etc; market-led land reform; climate mechanisms; microcredit
Approach to the Food Crisis	Increased industrial production; unregulated corporate monopolies; land grabs; expansion of GMOs; public-private partnerships; Liberal markets; microenterprise; Int'l. sourced food aid; The Global Agriculture and Food Security Program	Same as Neoliberal but w/ increased middle peasant production & some locally-sourced food aid; microcredit; more agricultural aid, but tied to GMOs & "bio-fortified/ climate-resistant" crops. Comprehensive Framework for Action (CFA), Feed the Future, Lets Move Campaign, Zero Hunger (Brazil)
Key documents	World Bank 2009 Development Report	World Bank 2009 Development Report

POLITICAL TRENDS		
FOOD MOVEMENTS		
POLITICS	**Progressive**	**Radical**
DISCOURSE	**Food Justice**	**Food Sovereignty**
Main Institutions	UN Committee on Food Security; COAG; We Are the Solution; ROPPA, PELUM; Groundswell, Campesino-a-Campesino, Slow Food, Community Food Security Coalition, CIW, Crossroads Center, many smaller philanthropic foundations and alternative fair trade	Via Campesina, EHNE, Friends of the Earth, MST, CLOC, COAG, NFFC, Food & Water Watch, ROPPA, ESAFF, African Food Sovereignty Alliance, People's Community Market, Food Chain Workers Alliance, ROC-United, Xarxa, Plataforma Rural, Food Sovereignty Alliance, European Climate Justice Action, Institute for Social Ecology, Community to Community, International Planning Committee on Food Sovereignty; World March of Women
ORIENTATION	Empowerment	Entitlement/Redistribution
MODEL	Agroecologically-produced local food; investment in under-served communities; new business models and community benefit packages for production, processing & retail; better wages for ag. workers; solidarity economies; land access; regulated markets & supply	Dismantle corporate agri-foods monopoly power; parity; redistributive land reform; community rights to water & seed; Regionally-based food systems; Democratization of food system; sustainable livelihoods; protection from dumping/ overproduction; Revival of agroecologically managed peasant agriculture to distribute wealth and cool the planet
Approach to the Food Crisis	Right to food; Better safety nets; sustainably-produced, locally-sourced food; agroecologically-based agricultural development; Committee on World Food Security (CFS)	Human right to food; Locally sourced, sustainably produced, culturally appropriate, democratically controlled; focus on UN/FAO negotiations
Key documents	IAASTD	Declaration of Nyelení; Peoples' comprehensive framework for action to eradicate hunger; ICARRD

and sustainable food systems. Both overlap significantly in their approaches. Together, folks in this global food movement seek to open up food systems to serve people of color, smallholders, and low-income communities while striving for sustainable and healthy environments. Radicals and progressives are the arms and legs of the same food movement.

Time for Transformation

The current food crisis reflects the environmental vulnerability, social inequity, and economic volatility of the corporate food regime. Clearly, without profound changes to the regime we will continue to experience cycles of free-market liberalization and mild regime reform, plunging the world's food systems into ever-graver crises. While food system reforms—such as localizing food assistance, increasing aid to agriculture in the Global South, increasing food stamps, and funding organic agricultural research—are certainly needed and long overdue, they don't alter the balance of power within the food system and, in some cases, may even reinforce the status quo.

Progressive projects are tremendously energetic, creative, and diverse, but can also be locally focused and issue rather than system driven. For example, the movement to improve access to food in low-income urban communities addresses a pressing need. But the causes of nutritional deficiency among underserved communities go beyond the location of grocery stores. The abysmal wages, unemployment, skewed patterns of ownership and inner-city blight, and the economic devastation that has been historically visited on these communities are the result of structural racism, globalization, and class struggles lost. No amount of fresh produce will fix urban America's food, and class can't be ignored or willed away. An honest and committed effort to the original food justice principles of antiracism and equity *within* the food movement is just as important as working for justice in the broader food system. Addressing the rights of women, labor, and immigrants is essential for strengthening movements for food justice. Rural-urban and North-South divides must also be addressed in practice and in policy for the food movement to unite in a significant way.

In this regard, the progressive trend of the food movement is pivotal: If progressive organizations build their primary alliances with reformist institutions from within the corporate food regime, the regime will be strengthened, and the food movement will be weakened. In this scenario, we are unlikely to see substantive changes to the status quo. However, if progressive and radical organizations find ways to build strategic alliances between them, the food movement will be strengthened. A united food movement has a much higher likelihood of pressuring legislators, bringing about reforms, and moving our food systems toward true transformation.

Another food system *is* possible; the political convergence of the world's food movements will bring it to life.

About the Authors

Samir Amin is director of the Third World Forum in Dakar, Senegal. He is an economist and internationally acclaimed author of over 30 books.

Eric Holt-Giménez is the Executive Director of Food First. He worked as an agroecologist for over 25 years with the Campesino a Campesino Movement. A researcher, writer and lecturer, his works address hunger, social movements and sustainable food systems.

Paul Nicholson is a member of the Basque Farmers' Union (EHNE—*Euskal Herriko Nekazarien Elkartasuna*) in the Basque Country, and a member of the International Coordinating Committee in *La Via Campesina.*

Horácio Martins de Carvalho is an agronomist, social scientist and consultant for La Via Campesina who looks at global food systems and Brazilian agriculture.

João Pedro Stedile is an economist and member of the coordinating body of the National MST and La Via Campesina, Brazil.

George Naylor, former president of the National Family Farm Coalition NFFC, farms grain on 470-acres in Churdan, Iowa. He is a graduate of the University of California, Berkeley (1971).

Tabara Ndiaye, from the Casamance region of Senegal, is a program consultant with The New Field Foundation. She works on building the capacity of rural women's associations in French-speaking West Africa.

Mariamé Ouattara, from Burkina Faso, is a program consultant to The New Field Foundation in the northern Niger River Basin. She is a founding member of REFAE, a regional network of African female economists that supports gender equality in macroeconomic politics.

John Wilson, free-range facilitator and resource person, is a small farmer in Zimbabwe. He helped establish Fambidzanai, a local NGO promoting ecological agriculture, and the regional PELUM Association.

Groundswell International, established in 2009, is a partnership of local NGOs and resource people in Latin America, Africa and Asia, coordinated by a global office based in the US. They work to strengthen capacity for positive social change in rural communities.

Raj Patel was a Policy Analyst with Food First from 2002 to 2004. He is the author of two popular books on our food and economic systems, *Stuffed and Starved*, and the recent *The Value of Nothing: How to reshape market society and redefine democracy.*

Josh Viertel is president of Slow Food USA. He previously co-founded and co-directed the Yale Sustainable Food Project at Yale University.

Brahm Ahmadi is co-founder and former Executive Director of People's Grocery. He recently left People's Grocery to launch a spin-off, startup venture called People's Community Market, which is developing a food retail model for inner city markets.

Lucas Benítez is a farmworker and one of the founders of the Coalition of Immokalee Workers (CIW) which has waged a successful campaign against abuses of immigrant workers in the US.

José Oliva is a restaurant worker and coordinator of the Restaurant Opportunities Center in Chicago, Illinois USA, a member of the Food Chain Workers Alliance.

Xavier Montagut, president of *Xarxa de Consum Solidari,* is an economist specializing in international trade, responsible consumption and fair trade. He is the co-author of several books including, *Supermarkets? No thank you.*

Ken Meter is president of Crossroads Resource Center. He has 39 years of experience in inner city and rural community capacity building.

Olivier De Schutter is the UN Special Rapporteur on the Right to Food. He teaches international human rights law, EU law and legal theory at the Catholic University of Louvain in Belgium, and the College of Europe and Columbia University.

Hans R. Herren, president of the Millennium Institute and winner of the 1995 World Food Prize, is a scientist who champions integrated sustainable development. He was co-chair of the International Assessment of Agricultural Knowledge, Science and Technology for Development.

Nora McKeon, formerly of the FAO, is a consultant and lecturer on food systems, peasant farmer movements and UN-civil society relations. She coordinates Terranuova, an exchange and advocacy program for African and European farmers' organizations.

Brian Tokar is an activist, author and critical voice for ecological activism since the 1980's. He is Director of the Institute for Social Ecology and is a lecturer in Environmental Studies at the University of Vermont.

Miriam Nobre is an agronomist, author and program coordinator of *Sempreviva Organização Feminista—SOF* (Evergreen Feminist Organization). She is part of the Economy and Feminism Network (*Rede Economia e Feminismo*) and part of the international coordination of the World March of Women.

Rosalinda Guillén, Executive Director of Community to Community Development in Bellingham, Washington, is a former farmworker. She organized with Cesar Chavez in the United Farmworkers Union.

Annie Shattuck is a Food First Fellow and graduate student at the University of California, Berkeley. She was a policy analyst at Food First from 2008-10 and co-authored *Food Rebellions!*

Resources

Food First

The purpose of the Institute for Food and Development Policy, otherwise known as Food First, is to eliminate the injustices that cause hunger. Programs: Building Local Agri-Foods Systems; Democratizing Development: Land, Resources and Markets; Forging Food Sovereignty with Farmers.
www.foodfirst.org

Basque Farmers Union, EHNE

EHNE is a farmers union and a professional agrarian organizations, formed and legalized in 1976-77. Presently it has 4 provincial members, EHNE Viscaya, EHNE Gipuzkoa, EHNE Navarra and UAGA of Alva. EHNE belongs to COAG at the national level, to CPE in Europe and La Via Campesina, internationally.

http://www.ehne.org/

Movimento dos Trabalhadores Rurais Sem Terra, MST

Brazil's Landless Worker's Movement
Twenty-six years ago in Cascavel (PR) hundreds of rural workers decided to found an autonomous peasant social movement to struggle for land, for Agrarian Reform and for the social transformations necessary for Brazil. They were squatters, displaced people, migrants, sharecroppers and peasant farmers; all landless rural workers who were deprived of their right to produce food. Since our founding, the Landless Workers Movement has organized based on three principal objectives: Struggle for land; Struggle for Agrarian Reform; Struggle for a more just and fraternal society. These objectives are manifested in our documents that orient the MST, defined in our National Congress and in our Program for Agrarian Reform.

http://www.mst.org.br/

Friends of the MST (USA): http://www.mstbrazil.org/

Journal of Peasant Studies

The Journal of Peasant Studies is one of the leading journals in the field of rural development. It was founded on the initiative of Terence J. Byres and its first editors were Byres, Charles Curwen and Teodor Shanin. It provokes and promotes critical thinking about social structures, institutions, actors and processes of change in and in relation to the rural world. It encourages inquiry into how agrarian power relations between classes and other social groups are created, understood, contested and transformed. The *Journal* pays special attention to questions of 'agency' of marginalized groups in agrarian societies, particularly their autonomy and

capacity to interpret – and change – their conditions.
www.tandf.co.uk

National Family Farm Coalition NFFC

The National Family Farm Coalition represents family farm and rural groups whose members face the challenge of the deepening economic recession in rural communities. The NFFC was founded in 1986.
http://www.nffc.net/

New Field Foundation

New Field Foundation contributes to the creation of a peaceful and equitable world by supporting women and their families to overcome poverty, violence and injustice in their communities.
http://www.newfieldfound.org/

Participatory Ecological Land Use Management PELUM

Participatory Ecological Land Use Management (PELUM) Association is a network of Civil Society Organizations / NGOs working with small-scale farmers in East, central and Southern Africa. The Association membership has grown from 25 pioneer members (in 1995) to over 230 members in 2010.
http://pelum.net/about-2.html

Eastern and Southern Africa Small Scale Farmers Forum

ESAFF is a network of small holder farmers that advocate for policy, practice and attitude change that reflects the needs, aspirations, and development of small-scale farmers in east and southern Africa. It was established in 2002 after the World Summit of Sustainable Development (WSSD) held in Johannesburg in South Africa.
http://www.esaff.org/

La Via Campesina

La Via Campesina is the international movement which brings together millions of peasants, small and medium-size farmers, landless people, women farmers, indigenous people, migrants and agricultural workers from around the world. It defends small-scale sustainable agriculture as a way to promote social justice and dignity. It strongly opposes corporate driven agriculture and transnational companies that are destroying people and nature. La Via Campesina comprises about 150 local and national organizations in 70 countries from Africa, Asia, Europe and the Americas. Altogether, it represents about 200 million farmers. It is an autonomous, pluralist and multicultural movement, independent from any political, economic or other type of affiliation.

http://viacampesina.org

Groundswell International

Groundswell International is a partnership of local NGOs and resource people in Latin America, Africa and Asia, coordinated by a global office based in the U.S., working to strengthen capacity for positive social change in rural communities. Groundswell is a tax-exempt, nonprofit 501(c)(3) organization.

http://groundswellinternational.org/

Boston Food Project

Boston Food Project's mission is to is to grow a thoughtful and productive community of youth and adults from diverse backgrounds who work together to build a sustainable food system. We produce healthy food for residents of the city and suburbs and provide youth leadership opportunities. Most importantly, we strive to inspire and support others to create change in their own communities.

http://thefoodproject.org/

Slow Food USA

Slow Food is an idea, a way of living and a way of eating. It is a global, grassroots movement with thousands of members around the world that links the pleasure of food with a commitment to community and the environment.
http://www.slowfoodusa.org/

People's Community Market

People's Community Market: For many years, People's Grocery worked toward opening a for-profit grocery store in West Oakland. We incubated and successfully spun off a project to build People's Community Market (PCM) in January of 2010, of which Brahm Ahmadi (co-founder of People's Grocery) is the founder and CEO. We will be one of PCM's partners when it opens its doors in 2012, contributing nutrition programming, leadership development, and providing pathways for the store to build relationships with residents.
http://www.peoplescommunitymarket.com/

People's Grocery is a health and wealth organization - our mission is to improve the health and economy of West Oakland through the local food system. We pursue positive community change and address social determinants of health through a food lens. We work to ensure that community self-determination plays a large part in the revitalization of low-income neighborhoods.
http://www.peoplesgrocery.org/

Crossroads Resource Center

A non-profit organization, works with communities and their allies to foster democracy and local self-determination. We specialize in devising new tools communities can use to create a more sustainable future.
http://www.crcworks.org/?submit=homepage

Coalition of Immokalee Workers, CIW

The CIW is a community-based organization of mainly Latino, Mayan Indian and Haitian immigrants working in low-wage jobs throughout the state of Florida. We strive to build our strength as a community on a basis of reflection and analysis, constant attention to coalition-building across ethnic divisions, and an ongoing investment in leadership development to help our members continually develop their skills in community education and organization.

From this basis we fight for, among other things: a fair wage for the work we do, more respect on the part of our bosses and the industries where we work, better and cheaper housing, stronger

laws and stronger enforcement against those who would violate workers' rights, the right to organize on our jobs without fear of retaliation, and an end to involuntary servitude in the fields.
www.ciw-online.org

ROC United

ROC-United is a national restaurant workers' organization that engages in six programs: 1) developing new restaurant worker organizing projects; 2) providing training and technical assistance to restaurant worker organizing projects; 3) conducting national research on the restaurant industry: 4) engaging in national policy work to improve working conditions for restaurant workers, including initiating and managing a national restaurant worker health insurance program; 5) coordinating national campaigns of restaurant workers; and 6) convening restaurant workers across the country.
www.rocunited.org

The Food Chain Workers Alliance, FCWA

The FCWA is a coalition of worker-based organizations whose members plant, harvest, process, pack, transport, prepare, serve, and sell food, organizing to improve wages and working conditions for all workers along the food chain. The Alliance was founded in July 2009. The Alliance works together to build a more sustainable food system that respects workers' rights, based on the principles of social, environmental and racial justice, in which everyone has access to healthy and affordable food.
http://foodchainworkers.org/?page_id=38

Xarxa Solidari

Xarxa de Consum Solidari (XCS) is an association created in 1996 working in Catalunya for fair trade and responsible consumption. Our integral vision for fair trade considers the whole food chain, from its production to its consumption. We defend fair trade with the commitment to radically transform the economic system, so all producers receive a fair price for their products, respect the environment and women rights, and produce healthy and culturally accepted food. We consider that producers and consumers should decide what products are produced. We defend food sovereignty of the people over the corporate industrialized food regime that dominates us. We work for fair trade and responsible consumption through developing cooperatives; educating and promoting critical action; responsible tourism; selling in shops, ecological cooperatives and fair trade business. www.xarxaconsum.net

Office of the UN Special Rapporteur on the Right to Food

The mandate of the Special Rapporteur is to promote the full realization of the right to food and the adoption of measures at the national, regional and international levels for the realization of the right of everyone to adequate food and the fundamental right of everyone to be free from hunger so as to be able fully to develop and maintain their physical and mental capacities;

The Special Rapporteur reports both to the UN General Assembly (Third Committee) and to the Human Rights Council on the fulfilment of the mandate. Contributions by all the stakeholders on the different issues mentioned above are welcomed. In addition to addressing structural issues threatening the full enjoyment of the right to food, the Special Rapporteur may send communications to governments, called letters of allegation, in urgent cases brought to his attention by reliable sources. A fuller description of the mandate of the Special Rapporteur can be found on the relevant page of the website of the Office of the High Commissioner for Human Rights.
http://www.srfood.org/

IAASTD

The objective of the International Assessment of Agricultural Knowledge, Science and Technology for Development (IAASTD) was to assess the impacts of past, present and future agricultural knowledge, science and technology on the: reduction of hunger and poverty,

improvement of rural livelihoods and human health, and equitable, socially, environmentally and economically sustainable development. The IAASTD was initiated in 2002 by the World Bank and the Food and Agriculture Organization of the United Nations (FAO) as a global consultative process to determine whether an international assessment of agricultural knowledge, science and technology was needed. The outputs from this assessment are a Global Assessment and five Sub-global Assessments; a Global and five Sub-Global Summaries for Decision Makers; and a cross-cutting Synthesis Report with an Executive Summary. The Summaries for Decision Makers and the Synthesis Report specifically provide options for action to governments, international agencies,
http://www.agassessment.org/

Institute for Social Ecology

Social ecology advocates a reconstructive and transformative outlook on social and environmental issues, and promotes a directly democratic, confederal politics. As a body of ideas, social ecology envisions a moral economy that moves beyond scarcity and hierarchy, toward a world that re-harmonizes human communities with the natural world, while celebrating diversity, creativity and freedom.
http://www.social-ecology.org/

World March of Women

The World March of Women was born in 2000 as a mobilization that brought together women from around the world in a campaign against poverty and violence. That year, the national coordinating bodies organized national demonstrations and other actions. The activities began on March 8th, the International Women's Day, and ended on October 17th, and were organized around the call: "2000 reasons to march against poverty and sexist violence." This action had a participation of over 6,000 groups in 161 countries and territories. Its closure activity mobilized thousands of women around the world. On this occasion, it was delivered to the United Nations (UN) in New York a document with 17 demands backed by 5 million signatures. http://www.mmf2010.info/news-1/tercera-accion-internacional-en-el-fsm-2011?set_language=en

Centro de Estudios sobre Movimientos Sociales; CEMS

Centro de Estudios sobre Movimientos Sociales (Center for Social Movement Studies) has five main objectives: To study social movements, their history and actual participation considering sociology, philosophy, political science, history, social psychology, cultural studies and communications; to ensure academic and action-based collaboration with people involved in social movements; to offer research support. Follow up what has been published in articles, thesis, etc. on social movements; and to share and spread the research information and observations through educational and popular activities.
http://www.upf.edu/moviments/es/

Community to Community Development

Community to Community Development is a women-led, place based, grassroots organization working for a just society and healthy communities. We are committed to systemic change and to creating strategic alliances that strengthen local and global movements towards social, economic and environmental justice.
http://foodjustice.org/

US Food Sovereignty Alliance

The US Food Sovereignty Alliance (USFSA) works to end poverty, rebuild local food economies, and assert democratic control over the food system. We believe all people have the right to healthy, culturally appropriate food, produced in an ecologically sound manner. As a US-based alliance of food justice, anti-hunger, labor, environmental, faith-based, and food producer groups, we uphold the right to food as a basic human right and work to connect our local and national struggles to the international movement for food sovereignty http://www.usfoodsovereigntyalliance.org/

Acronyms

AATF: African Agricultural Technology Foundation

AGRA: Alliance for a Green Revolution in Africa

AKST: Agricultural Knowledge, Science and Technology

AoA: Agreement on Agriculture

CAFTA–DR: Central American Free Trade Agreement (including the Dominican Republic)

CAP: Common Agricultural Policy (European Union)

CFA: Comprehensive Framework for Action

CFS: Committee on Food Security

CGIAR: Consultative Group on International Agricultural Research

CIMMYT: International Maize and Wheat Improvement Center

CIW: Coalition of Immokalee Workers

CSA: community supported agriculture

DFID: British Department for International Development

EU: European Union

FAO: Food and Agriculture Organization of the United Nations

FTA: free trade agreement

GATT: General Agreement on Tariffs and Trade

GDP: gross domestic product

GMO: genetically modified organism

HYV: high-yielding hybrid varieties

IAASTD: International Assessment of Agricultural Knowledge, Science and Technology for Development

ICARRD: International Conference on Agrarian Reform and Rural Development

IFC: International Finance Corporation

IFAD: International Fund for Agricultural Development

IFPRI: International Food Policy Research Institute

IISD: International Institute for Sustainable Development

IMF: International Monetary Fund

IPCC: United Nations Intergovernmental Panel on Climate Change

IRDP: integrated rural development projects

IRRI: International Rice Research Institute

LDCs: less developed countries

NAFTA: North American Free Trade Agreement

NGOs: non-governmental organizations

OECD: Organization for Economic Cooperation and Development

OPEC: Organization of the Petroleum Exporting Countries

RAI Responsible Agricultural Investment

SAP: structural adjustment program

UNDP: United Nations Development Program

UNCTAD: United Nations Conference on Trade and Development

USDA: US Department of Agriculture

WTO: World Trade Organization

Glossary

agroecology—the science of sustainable agriculture; a scientific discipline that uses ecological theory to study, design, manage and evaluate agricultural systems that are productive but also resource conserving. Agroecology links ecology, culture, economics, traditional knowledge and integrated management to sustain agricultural production and healthy food and farming systems.

agroforestry—a dynamic, ecologically based, natural resource management system that, through the integration of trees in farm and rangeland, diversifies and sustains production for increased social, economic and environmental benefits.

agrofuels—biologically-based fuels produced on a centralized, industrial scale, mostly for use as a liquid vehicle fuel. Agrofuels can be made from corn, soy, sugarcane, canola, jatropha, palm oil, or so-called "second generation" crops such as switchgrass, Miscanthus (canary grass), trees, and corn stover. The term contrasts with "biofuels," which refers to local, decentralized, farmer-owned, and small-scale fuels of a similar nature.

Archer Daniels Midland—ADM is the second largest grain trader in the world, a major food processor, and the second largest ethanol producer in the US. ADM has been called the "largest recipient of corporate welfare in U.S. history" by the conservative Cato Institute.

bushel—the unit of measurement in which corn and other commodities are most often traded. One bushel of corn = 56 pounds or 25.4 kg.

captive supply—describes the livestock that are committed to a specific buyer (meatpacker) two or more weeks in advance of slaughter. Captive supply operates where the market is dominated by a few firms. They use their market power to subvert the natural forces of market price determination by increasing production in order to depress the prices that farmers get for their slaughter livestock.

Cargill—the world's largest grain trader and the largest privately held company in the US.

commodity—a good for which there is demand, but which is supplied without qualitative differentiation across a market; the market treats it as equivalent no matter who produces it, e.g. corn.

community food security—a condition in which all community residents obtain a safe, culturally acceptable, nutritionally adequate diet through a sustainable food system that maximizes community self-reliance and social justice. ii

community grain banks—a localized system of storing food grain purchased during good harvests when prices are low. It provides a safety net for the community by making affordable grain available during difficult harvests and increased food prices.

corporate democracies—when a few corporations dominate the food system, and thus have the ability to influence social and economic policy. Accountability is not to citizens of the state, but to the corporations.

crop board—an independent government body that markets and regulates the price of crops.

Doha round—the current round of WTO negotiations, which began in 2001 in Doha, Qatar. The negotiations have stalled over disagreements on agricultural import rules.

dumping—export of overproduced and/or subsidized commodities, often from industrial Northern countries, distributed below the cost of production, most often in the global South.

emerging economy—used to describe a nation in the process of rapid industrial growth, such as China, India and Brazil.

food regime/corporate food regime—a food regime is a "rule-governed structure of production and consumption of food on a world scale." The first global food regime spanned the late 1800s through the Great Depression and linked food imports from Southern and American colonies to European industrial expansion. The second food regime reversed the flow of food from the Northern to the Southern Hemisphere to fuel Cold War industrialization in the Third World. Today's corporate food regime is characterized by the monopoly market power and mega-profits of agrifood corporations, globalized meat production, and growing links between food and fuel.

food crisis—a term to describe the 2008 and 2011 worldwide dramatic and rapid hikes in food prices and the on-going problems with food availability, price volatility, and the environmental challenges to food production that have caused increased food poverty and political instability in some countries of the Global South.

food justice—a movement that attempts to address hunger by addressing the underlining issues of racial and class disparity and the inequities in the food system that correlate to inequities in economic and political power.

food movement—collective term for the individuals and groups of food growers, sellers, processors and consumers working towards addressing the social and economic problems inherent in local, national, and global food systems.

food policy councils—a group of stakeholders that examine how the local food system is working and develop ways to fix it. Food policy councils are found at city, county and state levels.

food security—according to the FAO, "food security exists when all people, at all times, have physical and economic access to sufficient, safe and nutritious food to meet their dietary needs and food preferences for an active and healthy life."

food sovereignty—people's right to healthy and culturally appropriate food produced through ecologically sound and sustainable methods, and their right to define their own food and agriculture systems; the democratization of the food system in favor of the poor.

futures—standardized legal agreements to transact in a physical commodity at some designated future time.

Global South—formerly referred to as the "third world," the nations of Africa, Central and South America, and much of Asia with comparatively little economic power.

GMO—an acronym for genetically modified organism, a plant or animal with permanently, artificially modified genetic material derived across species boundaries. In reference to agriculture, this refers to proprietary, modified crop varieties.

Green Revolution—largely funded by the Ford and Rockefeller Foundations, the Green Revolution refers to the process of industrialization in agriculture initiated in the 1950s and 60s, from the development and widespread adoption of high-yielding varieties, synthetic fertilizers, chemical herbicides and pesticides.

genetic engineering—experimental or industrial technologies used to alter the genome of a living cell so that it can produce more or different molecules than it is already programmed to make.

grain reserves—a stock of grain maintained in years of good harvest to buffer against shortage and regulate price volatility.

hedging—a mechanism to offset the risk of an asset's changing price.

hypermarket—a large retailer that combines a supermarket and a department or general merchandise store under one roof. Hypermarkets like WalMart, Carrefour, Target, K-mart, and Costco, often covering 14,000m2 (150,000ft2), survive on high-volume, low margin sales, and often put local retailers out of business.

industrial agrifood complex—describes the skewed power structure of the global food system, currently dominated by large grain traders, chemical and biotechnology companies, transnational food processors, and global supermarket chains, at the expense of small farmers who produce most of the world's food.

index investors—type of speculator that seeks long-term investments by hoarding commodities futures contracts for extended periods and betting on the continued rise of commodities prices.

industrial feedlot—a type of a confined animal feeding operation, where animals are fattened on grains and soy before slaughter for meat.

informal sector—economic activity that is neither taxed nor monitored by the government.

industrial feedlot—a type of a confined animal feeding operation, where animals are fattened on grains and soy before slaughter for meat.

Intercrop or polyculture—a technique employed in traditional and ecological agriculture that involves the planting of multiple varieties and crop species in one agricultural area.

monocrop or monoculture—the practice of cultivating a single variety of genetically uniform plants over a large agricultural area.

Naylor Curve—used to describe the paradox in which farmers find themselves when prices drop, they begin to produce more, but this further degrades arable land even as increased volume in the market drives prices lower.

Neoliberalism—an approach to social and economic policy that promotes reduced state intervention, market liberalization and reduced regulations, free trade, and therefore seeks to maximize the power of the private sector.

public–private partnerships—a government service or business venture funded and managed jointly by government agencies and business.

smallholder—a farmer with relatively few planted acres that relies primarily on family labor.

sovereign wealth fund—a state-owned investment fund composed of financial assets such as stocks, bonds, real estate, or other financial instruments funded by foreign exchange assets. SWFs tend to have a higher tolerance for risk than traditional foreign exchange reserves.

smallholder or family farm—a farmer with relatively few planted acres that relies primarily on family labor, in contrast to factory farms owned by agribusinesses.

structural violence—a constraint on human potential due to political or economic forces. Sources of structural violence can include unequal access to resources, political power, education, food, and health care, as well as racism, sexism, religious discrimination, and other forms of oppression. Structural violence often leads to physical acts of violence.

Via Campesina—an international movement of peasant farmers' organizations that advocates for food sovereignty.

More books from Food First

Food Sovereignty: Reconnecting Food, Nature and Community

Edited by Annette Desmarais, Nettie Wiebe, and Hannah Wittman

Advocating a practical, radical change to the way much of our food system currently operates, this book argues that food sovereignty is the means to achieving a system that will provide for the food needs of all people while respecting the principles of environmental sustainability, local empowerment and agrarian citizenship.

Paperback $24.95

Food Rebellions: Crisis and the Hunger for Justice

Eric Holt-Giménez, Raj Patel and Annie Shattuck

Food Rebellions! contains up to date information about the current political and economic realities of our food systems. Anchored in political economy and an historical perspective, it is a valuable academic resource for understanding the root causes of hunger, growing inequality, the industrial agri-foods complex, and political unrest.

Paperback $19.95

Beyond the Fence: A Journey to the Roots of the Migration Crisis

Dori Stone

This book examines how U.S./Mexico policy affects families, farmers, and businesses on both sides of the border, exposing irretrievable losses, but also hopeful advances. Companion DVD, Caminos: The Immigrant's Trail, with study guide.

Paperback, $16.95

Agrofuels in the Americas

Edited by Rick Jonasse

This book takes a critical look at the recent expansion of the agrofuels industry in the U.S. and Latin America and its effect on hunger, labor rights, trade and the environment.

Paperback, $18.95

Alternatives to the Peace Corps: A Guide to Global Volunteer Opportunities, Twelfth Edition

Edited by Caitlin Hachmyer

Newly expanded and updated, this easy-to-use guidebook is the original resource for finding community-based, grassroots volunteer work—the kind of work that changes the world, one person at a time.

Paperback, $11.95

Campesino a Campesino: Voices from Latin America's Farmer to Farmer Movement for Sustainable Agriculture

Eric Holt-Giménez

The voices and stories of dozens of farmers are captured in this first written history of the farmer-to-farmer movement, which describes the social, political, economic and environmental circumstances that shape it.

Paperback, $19.95

Promised Land: Competing Visions of Agrarian Reform

Edited by Peter Rosset, Raj Patel and Michael Courville

Agrarian reform is back at the center of the national and rural development debate. The essays in this volume critically analyze a wide range of competing visions of land reform.

Paperback, $21.95

Sustainable Agriculture and Resistance: Transforming Food Production in Cuba

Edited by Fernando Funes, Luis García, Martin Bourque, Nilda Pérez and Peter Rosset

Unable to import food or farm chemicals and machines in the wake of the Soviet bloc's collapse and a tightening U.S. embargo, Cuba turned toward sustainable agriculture, organic farming, urban gardens and other techniques to secure its food supply. This book gives details of that remarkable achievement.

Paperback, $18.95

The Future in the Balance: Essays on Globalization and Resistance

Walden Bello. Edited with a preface by Anuradha Mittal

A collection of essays by global south activist and scholar Walden Bello on the myths of development as prescribed by the World Trade Organization and other institutions, and the possibility of another world based on fairness and justice.

Paperback, $13.95

Views from the South: The Effects of Globalization and the WTO on Third World Countries

Edited by Sarah Anderson
Foreword by Jerry Mander. Afterword by Anuradha Mittal

This rare collection of essays by activists and scholars from the global south describes, in pointed detail, the effects of the WTO and other Bretton Woods institutions.

Paperback, $12.95

Basta! Land and the Zapatista Rebellion in Chiapas, Third Edition

George A. Collier with Elizabeth Lowery-Quaratiello
Foreword by Peter Rosset

The classic on the Zapatistas in its third edition, including a preface by Rodolfo Stavenhagen.

Paperback, $16.95

We encourage you to buy Food First Books from your local independent bookseller; if they don't have them in stock, they can usually order them for you fast. To find an independent bookseller in your area, go to: www.booksense.com.

Food First books are also available through major online booksellers (Powell's, Amazon, and Barnes and Noble), and through the Food First website, www.foodfirst.org. You can also order direct from our distributor, Perseus Distribution, at (800) 343-4499. If you have trouble locating a Food First title, write, call, or e-mail us:

Food First
398 60th Street
Oakland, CA 94618-1212 USA
Tel: (510) 654-4400
Fax: (510) 654-4551
E-mail: foodfirst@foodfirst.org
Web: www.foodfirst.org
If you are a bookseller or other reseller, contact our distributor, Perseus Distribution, at (800) 343-4499, to order.

Films from Food First

The Greening of Cuba

Jaime Kibben

A profiling of Cuban farmers and scientists working to reinvent a sustainable agriculture based on ecological principles and local knowledge. DVD (In Spanish with English subtitles), $35.00

Caminos: The Immigrant's Trail

Juan Carlos Zaldivar

Stories of Mexican farmers who were driven off their land, forced to leave their families and risk their lives to seek work in the U.S.

DVD and Study Guide, $20.00

How to Join Food First

Private contributions and membership gifts fund the core of Food First/ Institute for Food and Development Policy's work. Each member strengthens Food First's efforts to change a hungry world. We invite you to join Food First.

As a member you will receive a 20 percent discount on all Food First books. You will also receive our quarterly publications, Food First *News and Views* and *Backgrounders*, providing information for action on current food and hunger issues in the United States and around the world. If you want to subscribe to our Internet newsletter, *People Putting Food First*, send us an e-mail at foodfirst@foodfirst.org. All contributions are tax deductible. You are also invited to give a gift membership to others interested in the fight to end hunger. www.foodfirst.org

To become a member or apply to become an intern, just call or visit ourwebsite: www.foodfirst.org.